AF566710

Astronomy and astronautics

Astronomy and astronautics:

an enthusiast's guide to books and periodicals

Andy Lusis

Mansell Publishing Limited
London

First published 1986 by Mansell Publishing Limited
(A subsidiary of The H.W. Wilson Company)
6 All Saints Street, London N1 9RL, England

British Library Cataloguing in Publication Data

Lusis, Andy
Astronomy and astronautics: an enthusiast's guide to books and periodicals.
1. Astronomy—Bibliography
I. Title
016.52 Z5151

ISBN 0-7201-1795-X

Printed in Great Britain by
Whitstable Litho Ltd., Whitstable, Kent.

Contents

Foreword

Large numbers of astronomical books are now published every year — indeed, it might be true to say 'every month'. They are of differing standards, and range from the general to the highly specialized. To keep track of them all is a well-nigh impossible task.

Therefore, a comprehensive index of books which are suited to the amateur and the more serious student is of great value. Andy Lusis has provided such a list, and it will be extremely useful to all those who have more than a passing interest in astronomy. To give just one example: what books about Halley's Comet have been produced since the comet came in sight once more, toward the end of 1982? Look in the relevant section, and you will find all the references you need.

I am sure that the book will be well received, and it can, of course, be regularly updated as more and more books appear.

Patrick Moore OBE, DSc(Hon), FRAS

Preface

As I write Comet Halley is coming ever closer, and a new book to mark (cash in on) its return seems to appear every day. I have tried to record as many of these works as possible; many of them include material which will remain of interest to someone long after the comet has departed. Despite the occasional explosion of books on a particular topic, like Halley's Comet, or the Space Shuttle, and the mass of basic introductory works, I hope to show in these pages what a great variety of astronomy and space books is available. Whatever the reader's previous level of knowledge there should be something to interest anyone who is stimulated to explore the literature.

I should like to thank everyone who helped in any way whatsoever during the compilation of this bibliography, but I should stress that any errors and omissions are my own responsibility (I would be interested to hear of any that readers might discover). My thanks to the staff of the library at the Rutherford Appleton Laboratory where this project began, to Jon Hutchins at the Royal Greenwich Observatory library, Peter Hingley at the Royal Astronomical Society library, the staff of the reading room at the British Library Lending Division, at Birmingham Polytechnic to the library staff who dealt with my numerous requests and Barry McNeil, and to all the periodical publishers who responded to my requests for information and sample issues. Special thanks to Guy Fennimore who provided much useful assistance, and to my mother and my brother for their invaluable help.

Of the many libraries visited in the course of compiling this bibliography the following proved particularly useful, some because of their convenient location rather than the strength of their astronomy and astronautics collections: Birmingham Public Library, British Library Lending Division, Leicester University Library, Nottingham Central Library, Nottingham University Library, Oxford Central Library, Radcliffe Science Library, Royal Astronomical Society Library, Royal Greenwich Observatory Library, Rutherford Appleton Laboratory Library, and Wolverhampton Polytechnic Library.

Introduction

Aims

This work attempts to provide a comprehensive, annotated list of books and periodicals on astronomy and astronautics in English, which are intended for anyone interested in these subjects below the professional or undergraduate level. It brings together material from a wide range of sources, thus creating a single reference source. I hope that it will be useful to a broad readership: for libraries that maintain astronomy and/or astronautics collections of whatever size (even among users of academic and specialist libraries there can be few who restrict themselves entirely to technical publications), to fill in gaps, for updating, and to find out about material beyond the library's collection; for amateur astronomers, astronomical clubs and societies, and space interest groups; for teachers below the undergraduate level, or at that level where non-science students are concerned, for identifying background material and course texts; and for reviewers of popular works who might like to check on, or remind themselves of similar works already published.

Rather than merely listing the books and periodicals I have provided annotations to most entries to indicate intended level of readership, content, and special features, and to allow some degree of comparison between works covering the same topic. To achieve this I personally inspected as many as possible of the items listed; for the rest the annotations have been written, where possible, on the basis of published reviews. However the elusive goal of comprehensiveness has not been achieved: some works I know to exist have been omitted because I could not find information sufficient to verify the nature of their content or their intended readership.

Scope

This bibliography is in some respects a successor to Robert Seal's *A guide to the literature of astronomy* (entry 24 in the present bibliography), which included some material for the general reader and the amateur astronomer. Seal's book covered material published before 1977, but did not include astronautics or life in the universe. It was updated by his and Sarah Martin's *A bibliography of astronomy 1970–1981* (Littleton, CO: Libraries Unlimited, 1982), but only for reference and review literature intended primarily for professional astronomers. Another work with some similarity to the present one is Jean F. Blashfield's *Aerospace bibliography* (entry 15), but its overlap is relatively small as it has far less coverage of astronomy, includes children's books and some technical works, lists American editions only, and does not include periodicals.

Subject

The term 'astronomy' is used in its broadest sense to refer to the science that deals with the universe beyond the Earth's atmosphere, while 'astronautics' is taken to be the science and technology (and wider implications of) space travel. One obvious relationship between the two arises when spacecraft are used to carry out investigation of celestial bodies. This bibliography covers only works that are entirely (or almost entirely) devoted to these subjects. However, where other subjects are treated in an astronomical context (for example, in archaeoastronomy, where archaeological techniques are used to reveal the astronomical practices of ancient peoples), the works in question are included. Where astronomy is central to multidisciplinary works, as in, for example, books on extraterrestrial life, then those works are also included. Books dealing with physics related to astronomy, and with the terrestrial atmosphere are included selectively.

To keep the bibliography to manageable proportions I have left out the following topics: time-keeping, including sundials; astronavigation (these topics are covered by Seal); pseudo-science; ufology (flying saucers); and fiction, but all of these are represented to a greater or lesser extent within the contents of some of the works that are included. In the case of astronautics, to reduce even further the overlapwith Blashfield *(see above)*, I have given less coverage to topics that are strongly represented there, namely US space policy, satellite communications, solar energy, resources for teachers and model rocketry.

Level of readership

The items are intended for a wide-ranging level of readership, from complete beginners to professionals seeking information at an introductory level outside their own speciality. This range of readership would thus include students and teachers, and, for purposes of background reading, physics and astronomy undergraduates. Material for amateur astronomers with a high degree of specialist knowledge is, in general, beyond the scope of this bibliography, but a number of quite technical works are included. Items specifically for children are not included, but younger readers with a strong interest in astronomy or astronautics should find much to suit them. Some items primarily for younger readers, but likely to have a wider appeal, are included. There is no clear boundary between specialist and non-specialist works, so a certain amount of subjective judgement was involved in selecting material for inclusion. In most, though by no means all cases, mathematical content beyond elementary algebra was the deciding factor in excluding a book. A number of scholarly but not excessively technical works are included.

Type of material

The bibliography includes only books and periodicals; such items as calendars, maps and charts, filmstrips, video discs and computer software, unless accompanying a book, are not included. Also omitted are newsletters of local astronomical and other societies, and ephemeral works — which means, in general, those not commercially available. Only material in English (including that translated into English) is included. Collections of articles or essays, and other works that are only partly about astronomy and/or astronautics, are not listed. Study guides and instructor's manuals which accompany textbooks are not given separate entries. The careful reader will, however, find small exceptions to most of these rules.

Date of material

The present work includes books on astronomy and astronautics published after 1976. Books on astronomy published before 1977 that have subsequently appeared in new editions, or have been reprinted since then, are included, as are some books published before 1977 that were not included by Seal (entry 24). Books on astronautics published between 1966 and 1977 are included selectively. There is no cut-off date for new works; books published in 1985 are included where possible. A few that have not appeared at the time of writing are also listed (some in the form of late inclusions only in the annotations for other entries) to make the work as nearly comprehensive and as up to date as possible. Only current periodicals are listed.

I have used the annotations to some entries to mention pre-1977 works that I feel may still be of some interest, but this work should not be taken as a definitive guide to worthwhile earlier books. They are found mostly in the section on the history of astronomy, which also includes one blanket entry for selected earlier general histories.

Arrangement and Content of Entries

The entries are listed according to a relatively simple subject scheme (as outlined in the contents list), which I devised especially for this bibliography. This arrangement is intended to reflect the number of works on a particular topic as well as the structure of the disciplines of astronomy and astronautics. But note that in dealing with material of a largely non-technical nature the emphasis on subject coverage differs from that of the professional literature: the emphasis here is toward general works and away from the more specialized areas.

Entries

All the works listed (except those mentioned only in the notes of other entries) are individually numbered. These numbers are used in all cross-references and the indexes. Under each subject heading the entries are alphabetically arranged by the surname of the first or only author (or by title if no author is named), except in a few cases where very closely related works would be separated or listed in the wrong order. The bibliographic description represents a balance between the need for simplicity and the desire to provide as much information as any user might expect to find.

Author

Each entry is introduced by the name of the author or editor (if one is named). If there is more than one author (or editor) all co-authors (or editors) are named. Others involved in the production of the book are named if they appear prominently on the title page. If the author's name does not appear prominently he or she is still named, but within square brackets. Editors of periodicals are not named, but some current editors of annual publications are named if they are reasonably well known.

Title

The full title of each work listed is given. If the titles of British and American editions differ then both are mentioned, the former followed by '(UK)', the latter by '(US)'. The same principle applies if the paperback title differs from that of the hardcover edition, except that the abbreviation '(pbk)' is used. But where the subtitle is the same it is given in full only the first time. If a previous edition of a work had a different title, the old title is mentioned in the notes.

Publishing information

Details of the most recent British and American (or other English-language) editions, in hardcover and paperback, are included as applicable. Note that many US editions are distributed in Britain, and many British editions in the United States. Book club editions are not listed. The information given is: place of publication (American towns and all but the largest cities may be distinguished by the presence of the two-letter zipcode state abbreviation); publisher's name, with any imprint used for a particular book noted in parentheses, for example Penguin (Pelican), Doubleday (Anchor Press); and the year of publication. To avoid misleading the reader as to how up to date a book is the original date of publication is noted if the book listed is a reprint or translation.

For periodicals the address is given in full, and the date of first publication is given regardless of whether or not the current publisher was responsible for the first issue.

Following the date is the ISBN (International Standard Book Number), or ISSN (International Standard Serial Number) in the case of periodicals. The ISBN or ISSN provides a unique identification for each edition, or periodical title, and is frequently used by librarians and booksellers. Hardcover publication is assumed unless otherwise stated. A paperback (abbreviated as 'pbk') is taken to be anything having soft covers.

Prices are not given because they are subject to change, and because a number of the books listed are out of print.

Description

A concise guide is given to the physical make-up and special features of each item listed. The pagination given is the number of pages to the end of the index, usually excluding those given roman numerals at the beginning of a book, but adding any unnumbered plates (in the form of this example: 220 + 16p). Illustrations and diagrams are assumed to be black and white or two-colour, unless otherwise indicated. The term 'diagrams' (abbreviated as 'diags') is used to represent all outline drawings. Note that in most cases I have not examined every edition of a book listed; there may be some variation in the details of British and American editions, as well as of hardcover and paperback editions.

Rather than merely noting that a particular work has a glossary, bibliography and index, I have supplemented this information with a rough guide to the length of these features, in the form: (1) = brief, (2) = standard (about three to ten pages in length, depending on page size; fewer for the bibliography), (3) = extensive. The term 'bibliography' (abbreviated as 'bibl') is used to encompass textual references, and notes, as well as lists of readings. An additional symbol '(x)' is used to indicate books in which the bibliographical references are scattered throughout the work.

A very rough guide to the size of a work is given for larger items by the presence of the letter 'Q' (= quarto), which is used loosely to indicate a height (or length in some cases) above about 24 centimetres.

Notes

Annotations indicate the intended readership where necessary, provide a brief outline of the content, and mention any notable features. Where an item was not available for inspection the entry is prefaced by an asterisk (*), in which case where notes appear they were adapted from published reviews or notices, the source of which is credited. Some of the notes for recent books (mid- to late 1985) were adapted from publisher's announcements, usually by modifying the superlatives. A few have been taken from *Scanning the skies*, the catalogue of Sky Publishing of Cambridge, Massachusetts, which provides reliable descriptions of its own publications and very brief notes on selected books from other publishers. Quotations that appear in the annotations for books that have been examined were taken from the books themselves. Where a statement appears to the effect that a book presents a survey of 'current knowledge', it means current at the time the book was written, which will have been some time before publication. The length of an entry is not necessarily a reflection of the value of the item concerned.

Abbreviations

AAS	American Astronautical Society
AAVSO	American Association of Variable Star Observers
ads	Advertisements
ASP	Astronomical Society of the Pacific
BAA	British Astronomical Association
BBC	British Broadcasting Corporation
bibl	bibliography
BIS	British Interplantary Society
col.	colour
Dept.	Department
diags	diagrams (all line drawings)
ed.	edition, editor, editors
EP	Educational Publication (NASA)
ESA	European Space Agency
FAS	Federation of Astronomical Societies
glos	glossary
HMSO	Her Majesty's Stationery Office
IAU	International Astronomical Union
ICBM	intercontinental ballistic missile
illus	illustrations
IR	infrared
ISBN	International Standard Book Number
ISSN	International Standard Serial Number
JAS	Junior Astronomical Society
JBAA	*Journal of the British Astronomical Association*
JPL	Jet Propulsion Laboratory
mag.	magnitude
MHz	megahertz
NASA	National Aeronautics and Space Administration
NGC	New General Catalogue
NSI	National Space Institute
p	page, pages
pbk	paperback
Q	quarto (large-sized)
QJRAS	*Quarterly Journal of the Royal Astronomical Society*

RA	right ascension
RAS	Royal Astronomical Society
rev.	revised
S&T	*Sky & Telescope*
SETI	Search for Extraterrestrial Intelligence
SF	science fiction
SP	Special Publication (NASA)
STS	Space Transportation System
trans.	translation, translator
US GPO	United States Government Printing Office
USSEA	United States Space Education Association
VLF	very low frequency
vol.	volume

1 General Astronomy

Reference Works

Dictionaries

1 Bonneau, B. Lee and Billy A. Smith
ASTRONOMY ILLUSTRATED (3rd ed.)
Dubuque, IA: Kendall/Hunt, 1980, ISBN 0-8403-2168-6 pbk.
248p; illus, diags. Index (2).

A dictionary of the more common astronomical terms for beginners. Arranged in groups by topic: constellations; atoms, light and spectra; the Solar System; stellar systems; and galactic systems. All the terms are illustrated, some with cartoons.

2 Hopkins, Jeanne
GLOSSARY OF ASTRONOMY AND ASTROPHYSICS (2nd ed.)
Chicago: University of Chicago Press, 1980, ISBN 0-226-35171-8. 196p.

Compiled by a copyeditor for the *Astrophysical Journal*, who writes that she was urged on by a desire to understand the terms used by authors, and so began a glossary of the most commonly used terms. Overall too technical for the beginner, even though many basic terms are included among the 2,500 or so entries. The entries are fairly short (four lines or less in most cases), except for those covering the planets which include basic numerical data. Does not cover astronautics, and there are no biographical entries. Includes cross references.

3 Illingworth, Valerie (ed.)
THE MACMILLAN DICTIONARY OF ASTRONOMY (2nd ed.)
London: Macmillan, 1985, ISBN 0-333-39062-8; 0-333-39243-4 pbk.
437p; diags.

Written by groups of specialists, intended for the general reader and as a convenient reference for specialists, this dictionary contains about 2,300 entries (over 250 new to this edition) covering the whole of astronomy and adding pertinent terms from physics and mathematics, and important terms from space exploration. Most entries give a clear, concise definition, sometimes with the aid of a diagram. Some entries are longer descriptive pieces over one page long, e.g. Apollo, craters, exobiology, Mars, pulsars, resolving power, and sunspot cycle. Includes many cross references, but no biographical entries.

The first edition also appeared under various other titles: *The Facts on File dictionary of astronomy* (US), *The Anchor dictionary of astronomy* (US pbk), and *A dictionary of astronomy* (UK pbk).

4 Maddison, Robert E.W., (Valerie Illingworth ed.)
A DICTIONARY OF ASTRONOMY
London: Hamlyn, 1980, ISBN 0-600-32996-8. 208p; illus, diags. Bibl (1).

Designed for the general reader, this work is more basic than (3), containing about 1,500 entries, very few of which are as much as one page long. Includes many biographical entries, and also covers observatories, artificial satellites and space probes. Clear definitions of basic terms, with descriptions of the planets, good diagrams, and a useful list of abbreviations.

5 Mitton, Jacqueline
KEY DEFINITIONS IN ASTRONOMY
London: Frederick Muller, 1980, ISBN 0-584-10565-7; 0-584-10547-9 pbk; Totowa, NJ: Littlefield, 1982, ISBN 0-8226-0375-6 pbk. 168p; illus, diags.

Contains about 230 selected terms and names, 'which are . . . most likely to be encountered in a magazine article or television programme, for instance'. Definitions range from a few lines to two or more pages for some, e.g. constellations and Jupiter. Includes cross references.

6 Moore, Patrick
THE A-Z OF ASTRONOMY (3rd ed.)
Glasgow: Fontana, 1978, ISBN 0-00-635472-6 pbk. 192p; diags.

Over 400 entries cover most aspects of astronomy, though slanted toward the more traditional areas: excludes astronautics, but includes entries for astronomers and observatories. Most entries are fairly brief, but there are longer, descriptive ones for such topics as quasars, supernovae, Moon and Sun. (First appeared in 1966 as *The amateur astronomer's glossary*).

7 Nicolson, Iain
ASTRONOMY: A DICTIONARY OF SPACE AND THE UNIVERSE (UK)/A DICTIONARY OF ASTRONOMY (US)
London: Arrow, 1977, ISBN 0-09-914580-4 pbk; New York: Barnes &

Noble, 1980, ISBN 0-06-463524-4 pbk. 250 + 16p; diags. Bibl (2).

Begins with a brief introduction on the nature of astronomy. Contains about 500 entries, with coverage extending to space exploration, and physics where relevant. The length of entries ranges from a few lines to about two pages, though cosmology merits four.

8 Schewe, Philip F.
GLOSSARY OF TERMS USED IN COSMOLOGY
New York: American Institute of Physics, 1982, pbk. 23p; Bibl (1).

About 120 entries provide definitions based on the present state of knowledge, often giving a brief explanation rather than a simple definition. Includes cross references.

9 Tver, David F.
DICTIONARY OF ASTRONOMY, SPACE AND ATMOSPHERIC PHENOMENA
New York: Van Nostrand, 1979, ISBN 0-442-24045-7. 281p; diags.

Less comprehensive than (3), this work contains about 2,000 definitions and cross references clearly laid out. It is aimed at astronomers, science editors and writers, as well as lay readers. (Compiled with the assistance of Lloyd Motz and William K. Hartmann).

Encyclopaedias

*10 Angelo, Joseph A.
THE EXTRATERRESTRIAL ENCYCLOPEDIA
New York and Oxford: Facts on File, 1985, ISBN 0-87196-764-2. 256p; illus, Index. Q

11 Audouze, Jean and Guy Israël (general editors)
THE CAMBRIDGE ATLAS OF ASTRONOMY
Cambridge: Cambridge University Press, 1985, ISBN 0-521-26369-7 and Feltham: Newnes, 1985, ISBN 0-600-35814-3. 432p; illus & diags (mostly col.). Bibl (2), glos (3), index (3). Q

'A comprehensive account of the results of mankind's physical and intellectual exploration of the universe', suitable for anyone with a serious interest in astronomy. Includes contributions from a team of 44 specialists, mostly French, some British. This huge volume contains many excellent illustrations, but the type is rather small. The glossary includes some quite detailed explanations.

12 Mitton, Simon
THE CAMBRIDGE ENCYCLOPAEDIA OF ASTRONOMY
London: Jonathan Cape, 1977, ISBN 0-224-01418-8; New York: Crown, 1977, ISBN 0-517-52806-1. 495p; illus & diags (some col.). Glos (2), index (3). Q

'A broadly-based survey with an emphasis on firmly established new results', arranged in sections covering such topics as a survey of the universe, the stars, the Solar System, between the stars, life in the universe, and astronomy in space. The extensive cross referencing requires use of the index. Includes a 14-page star atlas and a nine-page outline of physics, with a glossary of physical terms.

13 Parker, Sybil P. (ed.), George O. Abell (consulting ed.)
McGRAW-HILL ENCYCLOPEDIA OF ASTRONOMY
New York: McGraw-Hill, 1983, ISBN 0-07-045251-2. 450 + 13p; illus (some col.), diags. Bibl (x), index (3). Q

A selection of 230 entries from the *McGraw-Hill encyclopedia of science and technology* (5th ed. 1982), alphabetically arranged, with cross references. Fairly detailed coverage of theoretical, observational and experimental aspects of astronomy and the instruments used by astronomers. No biographical entries. The level varies from elementary to quite technical.

14 Ridpath, Ian (ed.)
THE ILLUSTRATED ENCYCLOPEDIA OF ASTRONOMY & SPACE (2nd ed.)
London: Macmillan, 1979, ISBN 0-333-27653-1; New York: Crowell, 1980, ISBN 0-690-01838-X. 240 + 8p; illus (some col.), diags. Index (2). Q

A popular reference work, alphabetically arranged. The contributors are mostly well-known astronomy popularizers, such as Heather Couper, Paul Davies, Nigel Henbest and Colin Ronan. Coverage is broad rather than in-depth: the approximately 1,500 entries include biographies, and vary in length from brief, almost dictionary definitions to longer one to two-page summaries, e.g. Apollo programme, galaxies, life in the universe, rockets, and Venus. Includes cross references.

Others

15 Blashfield, Jean F.
AEROSPACE BIBLIOGRAPHY (7th ed.)
Washington, DC: NASA, 1982, pbk. 141p. Index (3).

The first new edition since 1972 provides an annotated list of selected books and reference materials (not periodicals) published 1971–80. Covers astronautics, communications, astronomy, and other fields encompassed by the work of NASA. Intended for teachers and general readers, it lists only publications commercially available in the US, but for all levels of readership below the specialist level (the entries are graded by reading level). The entries are listed first by subject, then, with annotation, alphabetically by author.

16 Chamberlain, Von Del (ed.)
SKY INTERPRETATION RESOURCE BULLETIN, VOL. V
Washington, DC: National Air & Space Museum, 1982, pbk. 80p; illus, diags. Bibl (3). Q

This volume largely supersedes previous volumes, which are out of print (and were published by the American Astronomical Society). A compendium of information for educators, it covers sky mythology and lore, and atmospheric phenomena as well as astronomy. Includes articles on Olbers' paradox, the Milky Way, astronomically inspired music, "A short bibliography and suggested photographic techniques on meteorological optics", star maps, and a good annotated bibliography. (Available free on request from the editor, Space Science & Exploration Dept., National Air & Space Museum, Smithsonian Institution, Washington, DC 20560, USA).

*17 Flick, T. Michael, (and others)
GUIDELINES FOR ASTRONOMY COURSES (2nd ed.)
Carpentersville, IL: Astronomical League, 1978, pbk. 68p.

Advice to amateur clubs for organizing, promoting, and funding a course in introductory astronomy, and aid for would-be instructors in designing effective classes. Includes descriptions of nearly 40 textbooks, and suggests plans to accompany four of them. (From *S&T* March 1979, p 289).

*18 Fraknoi, Andrew
UNIVERSE IN THE CLASSROOM: A RESOURCE GUIDE FOR TEACHING ASTRONOMY
New York: Freeman, 1985, ISBN 0-7167-1692-5 pbk. 269p.

A guide to books, articles and visual aids of use to classroom instructors in teaching astronomy. It is designed to accompany William Kaufmann's textbook *Universe* (91). Contains hints for teaching some of the concepts and provides sample multiple-choice and essay examination questions. (From *Astronomy* July 1985, p 57 and *S&T* Sept. 1985, p 229).

This book is probably a successor to the author's *Resource book for the teaching of astronomy* (San Francisco: Freeman, 1977, ISBN 0-7167-0288-6 pbk), which accompanied Ivan King's *The universe unfolding* (Freeman, 1976).

*19 Friedman, Alan J., Lawrence F. Lowery, Steven Pulos, Dennis Schatz and Cary I. Sneider
PLANETARIUM EDUCATOR'S WORKSHOP GUIDE
Rochester, NY: International Planetarium Society, 1980, pbk. 196p.

A collection of projects, activities and workshops devoted to the teaching of astronomy both in and out of a planetarium, with emphasis on student participation. (From *S&T* July 1981, p 60).

20 Gall, James
ASTRONOMICAL DIRECTORY
Toronto: Gall Publications, 1978, ISBN 0-88904-082-6 pbk. 144p. Q.

A worldwide listing of observatories, amateur groups, planetaria, suppliers, publications and publishers, and related institutions, produced from a computer printout. The main listing of all entries is by country (and by state for the US). Usually gives an address, sometimes supplemented by other information, such as number of members, projector type, and so on, as appropriate.

21 Jones, Brian (ed.)
HANDBOOK FOR ASTRONOMICAL SOCIETIES
Bradford: Federation of Astronomical Societies, annual, pbk. About 70p; illus, diags.

A compendium of general articles and reference information. Includes an index to articles in previous editions, details of the FAS and other astronomical organizations, periodicals, equipment suppliers, places to visit, speakers, overseas information sources, and visual aid sources.

22 NASA
RECORDS OF ACHIEVEMENT: NASA SPECIAL PUBLICATIONS
Washington, DC: NASA, 1983, NASA SP-470, pbk. 127p; col. illus, Index (2).

An updated version of NASA SP-449 *A catalog of NASA Special Publications* (NASA, 1981), which adds an index and a summary of NASA achievements. The introduction by Kay E. Voglewede gives a brief history of NASA SPs, those publications which do not conveniently fit into other categories. Many works in the General and History sections of the SP series are intended for the general public, and cover astronomy and space topics. Most such works are included in this bibliography. SP-470 lists about 1,200 SPs, and over 100 CPs (Conference Publications) and RPs (Reference Publications), with some brief annotation and details of US availability. Does not cover EPs (Educational Publications).

A list of astronomy-related SPs, and some other NASA publications, extracted from SP-449, appeared in *Mercury* (Sept./ Oct. 1981, p 152–58).

23 Schatz, Dennis, Andrew Fraknoi, Robert R. Robbins and Charles D. Smith
EFFECTIVE ASTRONOMY TEACHING AND STUDENT REASONING ABILITY
Berkeley, CA: Lawrence Hall of Science, University of California, 1978. 205p; illus, diags. Bibl (x).

Based on a workshop on teaching strategies for astronomy educators, this book is intended to help improve teaching strategies of introductory astronomy, especially of non-science students. The study incorporates the ideas of Jean Piaget who showed that, as they grow, students develop different kinds of reasoning ability. Includes readings on Piaget's theories, and a resource guide. The book also acts as a guide for planning similar workshops.

24 Seal, Robert A.
A GUIDE TO THE LITERATURE OF ASTRONOMY
Littleton, CO: Libraries Unlimited, 1977, ISBN 0-87287-142-8. 306p. Indexes (3).

A bibliography of selected books and periodicals which are specifically about astronomy: excludes astronautics. Contains 578 numbered entries covering astronomy at all levels, arranged by topic within four broad categories: reference sources, general materials, descriptive astronomy, and special topics (space science and astrodynamics, radio astronomy, the new astronomies). The entries provide a good description of the contents of each work, with some evaluation, about 60 to 200 words in length. (*See* Introduction for further details).

Introductory Works

Pictorial

The books listed here emphasize illustrative matter, but this does not necessarily imply that their text is insignificant. (*See also* 151–156).

25 Bergamini, David
THE UNIVERSE (2nd ed.)
Alexandria, VA: Time-Life, 1977, ISBN 0-7054-0105-7. 192p; illus (mostly col.). Q

A profusely illustrated general account of astronomy, with emphasis on the universe beyond the Solar System and cosmology. Suitable for younger readers.

26 Dixon, Don
UNIVERSE
Boston: Houghton Mifflin, 1981, ISBN 0-395-31290-6. 240p; illus (mostly col.), diags. Q

Largely a showcase for the work of the author, who is an astronomical artist. Contains many excellent colour reproductions of his work, and a number of photographs, which adorn this survey of our knowledge of the universe from the Solar System outwards. Includes 17 pages of planetary quantities compiled by T.A. Heppenheimer.

27 Feldman, Anthony
SECRETS OF SPACE (UK)/SPACE (US)
London: Aldus, 1980, ISBN 0-490-00462-8; New York: Facts on File, 1980, ISBN 0-87196-416-3. 336p; illus & diags (mostly col.). Index (3). Q

A profusely illustrated 'book-club' style volume, which gives a sketchy account of astronomical discoveries and of space exploration. Includes a chapter on UFOs.

28 Friedman, Herbert
THE AMAZING UNIVERSE
Washington, DC: National Geographic Society, 1980, ISBN 0-87044-179-5 (reprint of 1st ed. 1975). 200p; illus & diags (mostly col.). Glos (1), bibl (1), index (2).

A lavishly illustrated review of 'current' astronomy for the reader with no previous knowledge. The eight chapters describe not only the work of astronomers, but also convey the love of the subject which makes it more than just a job. Concentrates on extra-solar work, with a chapter on SETI.

29 Gallant, Roy A.
NATIONAL GEOGRAPHIC PICTURE ATLAS OF THE UNIVERSE
Washington, DC: National Geographic Society, 1980, ISBN 0-87044-356-9. 276p; illus & diags (mostly col.). Glos (1), index (1). Q

A book lavishly illustrated, but mostly with artwork rather than photographs. The undemanding, often simplistic text gives a brief introduction to astronomy and space exploration, ending with a speculative chapter on the future of space travel, which completes the rather unrealistic overall impression. The book is not very strongly bound.

30 Henbest, Nigel and Michael Marten
THE NEW ASTRONOMY
Cambridge: Cambridge University Press, 1983, ISBN 0-521-25683-6. 240p; illus & diags (mostly col.). Index (2). Q

A survey of modern astronomy, which describes the techniques for investigating the universe in all parts of the electro-magnetic spectrum now accessible to the astronomer, and the great increase in our knowledge which has resulted. However, this is largely a showcase for the 275 colour images reproduced here, some of which were specially processed for this volume.

31 Kerrod, Robin
STARS AND PLANETS (2nd ed.)
London: Ward Lock, 1983, ISBN 0-86272-080-X; 0-86272-062-1 pbk; New York: Arco, 1984, ISBN 0-668-06263-0 pbk. 125p; illus & diags (mostly col.). Glos (1), index (2).

A brief pocket guide, profusely illustrated in colour. It is easily readable and suitable for complete beginners of virtually any age. The main text is broken up by many short descriptions of features such as stars, lunar craters, and interstellar molecules. Information is sometimes presented in the form of short glossaries, e.g. features of the Sun. A bibliography would have been a useful addition.

32 Mitchell, James (ed.)
THE ILLUSTRATED REFERENCE BOOK OF THE UNIVERSE
Leicester: Windward, 1982, ISBN 0-85533-271-9 (reprint of 1st ed. 1980). 130p; illus & diags (mostly col.). Bibl (1), index (2). Q

Covers the whole of astronomy, with space exploration in uniform two-page spreads of which about a quarter is text (about 850 words). (Originally titled *Exploring the universe*).

33 Moore, Patrick
THE NEW ATLAS OF THE UNIVERSE (3rd ed.)
London: Mitchell Beazley, 1984, ISBN 0-85533-537-8; New York: Crown, 1984, ISBN 0-517-55500-X. 271p; illus & diags (mostly col.). Glos (3), index (3). Q

A large-sized, well illustrated atlas showing what is known about the universe, and guiding the novice to a better appreciation of it. The five sections cover history, the Solar System, beyond the Solar System, the future of astronomical discovery, ending with a guide to the sky for the naked-eye observer which includes star maps for the Northern and Southern skies. (Previously: *The atlas of the universe*).

34 Moore, Patrick
THE MITCHELL BEAZLEY NEW CONCISE ATLAS OF THE UNIVERSE (UK)/THE NEW CONCISE ATLAS OF THE UNIVERSE (US)
London: Mitchell Beazley, 1978, ISBN 0-85533-137-2; Chicago: Rand McNally, 1978, ISBN 0-528-83031-7. 190p; illus & diags (mostly col.). Glos (2), index (2). Q

A shorter version of (33).

35 Moore, Patrick
TRAVELLERS IN SPACE AND TIME
London: Park Lane Press, 1984, pbk; New York: Doubleday, 1984, ISBN 0-385-19051-4 pbk. 192p; illus & diags (mostly col.). Glos (1), index (2). Q

A profusely illustrated 'supermarket' book, which nevertheless provides a good introduction to astronomical knowledge for the complete beginner in the form of an imaginary journey from the Earth outwards.

36 Nicolson, Iain
ASTRONOMY
London: Hamlyn, 1982, ISBN 0-600-35653-1 (reprint of 1st ed. 1970). 160p; illus & diags (mostly col.). Bibl (1), index (1).

This colourful basic introduction includes sections on professional equipment and techniques, and amateur astronomy. Provides basic star charts, with notes on constellations. The bibliography has been updated only to 1972.

37 Ronan, Colin
DEEP SPACE
London: Macmillan, 1982, ISBN 0-333-34262-3; New York: Macmillan, 1982, ISBN 0-02-604510-9; London: Pan, 1983, ISBN 0-330-26906-2 pbk. 208p; illus & diags (some col.). Glos (1), index (2). Q

A companion volume to Ronan's *The practical astronomer* (180), which employs the same format: a large number of short sections giving an outline of astronomy, including cosmology, space probes, and the question of life in the universe. Includes a number of spectacular illustrations.

38 Warner, Lionel
STARS AND PLANETS OF THE SOUTHERN HEMISPHERE
Wellington, New Zealand: A.H. & A.W. Reed, 1978, ISBN 0-589-01036-0. 64 + 4p; illus (some col.) diags. Index (1).

A highly illustrated beginner's guide, with many explanatory diagrams and a concise, readable text. Concentrates on the Solar System, but also covers observing techniques with binoculars and small telescopes. Includes basic star charts.

Others

39 Alter, Dinsmore, Clarence H. Cleminshaw and John G. Phillips
PICTORIAL ASTRONOMY (5th ed.)
New York: Harper & Row, 1983, ISBN 0-06-181019-3. 374p; illus, diags. Glos (2), index (3). Q

Intended primarily for the lay reader or amateur astronomer, but 'numerous colleges and universities have adopted earlier editions for their "cultural" courses'. Virtually a textbook but with greater emphasis on illustrative matter. This edition has been extensively rewritten to include recent discoveries. Nearly 200 pages are devoted to the Solar System, over 100 pages cover stars and nebulae, while a shorter miscellaneous section covers such topics as life in the universe and man in space. The glossary has 'explanations more detailed than usual'. A drawback is the lack of a bibliography.

40 Asimov, Isaac
THE UNIVERSE: FROM FLAT EARTH TO QUASAR (3rd ed.)
New York: Walker, 1980, ISBN 0-8027-0655-X; Harmondsworth: Penguin, 1983, ISBN 0-14-021212-4 pbk. 408 + 16p; illus, diags. Bibl (1), index (3).

A beginner's guide from one of the best and certainly the most prolific of science popularizers. This is virtually a history of astronomical discovery, beginning with the Earth and moving outwards, giving clear, concise explanations.

41 Buedeler, Werner, (Fred Bradley, trans.)
THE FASCINATING UNIVERSE: THE MODERN ASPECTS OF ASTRONOMY
New York: Van Nostrand, 1983, ISBN 0-442-21427-8 (trans. of 1st German ed. 1981). 252 + 8p; illus (some col.), diags. Bibl (1), index (2).

A descriptive account of modern astronomy, based on the author's German TV series. Begins with a historical section to put what follows into perspective. Covers the Solar System, the Sun, and in lesser detail, the stars, and 'the universe in space and time'. The bibliography contains mostly German references.

42 Cattermole, Peter
ASTRONOMY
Wakefield: EP Publishing, 1978, ISBN 0-7158-0470-7. 96p; illus, diags. Bibl (1), index (1).

A brief, well illustrated (though not in colour) outline of astronomy, suitable for beginners of all ages.

43 Dutton, A.M.
JOURNEY TO THE STARS
Bognor Regis: New Horizon, 1978, ISBN 0-86116-013-4. 107p; illus, diags. Glos (2), index (1).

This low-budget production presents a general introduction to astronomy, concentrating mainly on the Solar System. Includes a seven-page who's who. The poorly reproduced illustrations are found together at the back of the book.

44 Evans, David S.
ASTRONOMY (rev. ed.)
London: Hodder & Stoughton (Teach Yourself Books), 1980, ISBN 0-340-15248-6 pbk (reprint of rev. ed. 1975). 290 + 16p; illus, diags. Glos (2), bibl (2), index (3).

Assumes no previous knowledge, but builds on that gained unconsciously to build up a basic knowledge of astronomy which is strongly slanted towards the traditional aspects of recognition, observation and description, as the basis of practical work. Modern aspects are covered briefly.

45 Knox, Richard
FOUNDATIONS OF ASTRONOMY: FROM BIG BANG TO BLACK HOLES
Newton Abbot: David & Charles, 1979, ISBN 0-7153-7660-8; New

York: Halsted Press, 1979, ISBN 0-470-26638-4. 184p; illus, diags. Bibl (1), index (2).

A rather brief introduction for the 'lay or student reader'. Coverage includes the history and development of astronomy and cosmology. 'Concentrates on the ways in which the various bodies of the universe function and interact'. An appendix describes how to make and use a simple telescope.

46 Lancaster-Brown, Peter
ASTRONOMY IN COLOUR (rev. ed.)
Poole: Blandford, 1979, ISBN 0-7137-0729-1 (reprint of rev. ed. 1975). 264p; col. illus, diags. Glos (1), index (2).

A general introduction, with a new foreword written in 1978, but otherwise not updated. The colour illustrations are lumped together in an 80-page mass in the centre of the book.

47 Maffei, Paolo, (D.J.K. O'Connell, trans.)
BEYOND THE MOON
Cambridge, MA: MIT Press, 1978, ISBN 0-262-13133-1; New York: Avon, 1980, ISBN 0-380-48744-6 pbk (trans. of 6th Italian ed. 1977). 377p; illus, diags, Index (3).

Closer in nature to a textbook than similar works, this book takes the form of an imaginary journey from the Moon outward. Concentrates on recent discoveries, with little historical background. Clearly very popular in Italy, with six editions in five years. (*See also* 121).

48 McAleer, Neil
THE COSMIC MIND-BOGGLING BOOK
New York: Warner, 1982, ISBN 0-446-37932-8 pbk; London: Hodder & Stoughton, 1983, ISBN 0-340-33429-0; 0-340-33428-2 pbk. 207p; illus, diags. Bibl (1), index (2).

A survey of astronomy, concentrating on recent findings presented in the style suggested by the title. The seven chapters, four of which cover the Solar System, take the reader on a 'gosh! wow!' tour of the universe, in short, snappy paragraphs each concentrating on a different feature.

49 Mitton, Jacqueline
ASTRONOMY: AN INTRODUCTION (UK)/ASTRONOMY: AN INTRODUCTION FOR THE AMATEUR ASTRONOMER (US)
London: Faber, 1978, ISBN 0-571-11232-3; 0-571-11252-4 pbk; New York: Scribner's, 1979, ISBN 0-684-16042-0. 148p; illus, diags. Bibl

(1), index (1).

A brief general introduction which 'grew out of notes prepared for a one-year school course leading to O level astronomy'. The UK edition includes O level standard questions.

50 Mitton, Jacqueline and Simon
DISCOVERING ASTRONOMY (UK)/THE PRENTICE-HALL CONCISE BOOK OF ASTRONOMY (US)
London: Longman, 1979, ISBN 0-582-39020-6 (reprinted 1983, ISBN 0-582-39226-8); Englewood Cliffs, NJ: Prentice-Hall, 1979, ISBN 0-13-166967-2. 96p; illus & diags (mostly col.). Index (1). Q

Intended primarily for younger readers, but suitable for adults lacking the confidence to tackle anything more difficult. Covers the whole of astronomy, with about two-thirds of the book devoted to the Solar System and observation. A bibliography would have been a useful addition.

51 Moore, Patrick
BASIC ASTRONOMY (2nd ed.)
Hornchurch: Ian Henry, 1979, ISBN 0-86025-823-8. 143p; illus, diags.

A brief outline covering the whole field of astronomy, which may prove too sketchy for readers with some previous knowledge.

52 Moore, Patrick
THE UNFOLDING UNIVERSE
London: Michael Joseph, 1982, ISBN 0-7181-2152-X; New York: Crown, 1982, ISBN 0-517-54836-4. 256p; col. illus. Index (2).

A book commemorating the 25th anniversary of the BBC TV series *The sky at night*. It is a personal account of the development of recent astronomy by the programme's well known presenter. The text is relatively short, with many excellent illustrations interspersed. Covers the whole of astronomy, including a summary of the historical background.

53 Murdin, Paul, David Allen and David Malin
CATALOGUE OF THE UNIVERSE
Cambridge: Cambridge University Press, 1979, ISBN 0-521-22859-X; New York: Crown, 1979, ISBN 0-517-53616-1. 256 + 16p; illus (some col.), diags. Glos (1), index (1). Q

Reverses the usual principle of astronomy books by illustrating 'a comprehensive selection of objects in the Universe and [writing] about the examples chosen, bringing the basic principles out of the objects themselves'. Arranged in three sections: the universe of galaxies, stars and nebulae, and the Solar System. An appendix lists the Messier objects.

54 Rowan-Robinson, Michael
COSMIC LANDSCAPE: VOYAGES BACK ALONG THE PHOTON'S TRACK
Oxford: Oxford University Press, 1979, ISBN 0-19-857553-X. 149p. Glos (2), index (1).

An interesting and unusual way of presenting the discoveries of astronomy, particularly the new branches, to the general reader. In six 'voyages' the author invites the reader to imagine a journey in which the universe is perceived in different bands of the electromagnetic spectrum: visible, radio, ultra-violet, x-ray and gamma-ray, infra-red, and microwave. The final chapter covers life on Earth and elsewhere. Suffers a little from the absence of illustrations.

55 Sidgwick, J.B., (rev. R.C. Gamble)
INTRODUCING ASTRONOMY (2nd ed.)
London: Faber, 1979, ISBN 0-571-04823-4 pbk (reprint of 2nd ed. 1973). 259p; diags. Glos (2), index (2).

A very traditional approach for potential observers. Includes a long (123 pages) appendix describing the constellations, with finder maps.

56 Thacker, Jerrold
THE DECEPTIVE UNIVERSE
Dallas, TX: ASTRO Press, 1982, ISBN 0-9608568-0-3. 216p; illus, diags. Bibl (1), index (2).

A general account which concentrates on physical principles and on describing recent findings beyond the visible part of the spectrum.

Textbooks

With such a large number of textbooks on the market there is bound to be a degree of similarity between them. The majority of the books listed below are aimed at the American non-science undergraduate, or liberal arts student, but many authors express the hope that the interested lay reader will find their books useful. Most of the texts follow the traditional pattern in their arrangement of topics; this means that they begin with a general survey of astronomy, then move on to discuss physical principles, or provide a historical background. They then cover the Solar System, stars and nebulae, stellar evolution, the Galaxy, extragalactic astronomy, and cosmology. Where significant variations occur they are noted. Many of the books use two-colour rather than black and white diagrams. The majority include several appendices which tabulate various astronomical data, list mathematical constants, and provide other supplementary information. Unless otherwise noted the texts employ virtually no mathematics. To provide practical experience for the students the texts may be supplemented by any of the numerous laboratory manuals available (*see* 299–313).

Basic/short course

57 Berg, Rebecca M. and Laurence W. Fredrick
DESCRIPTIVE ASTRONOMY
New York: Van Nostrand, 1978, ISBN 0-442-25472-5. 323 + 8p; illus (some col.), diags. Bibl (2), glos (2), index (2). Q

This text has a stronger historical emphasis than most, which is reflected in its arrangement, concentrating first on observed features, from the planets outward, then returning to concentrate on physical properties, formation and evolution. Ends with cosmology and life in the universe. Includes chapter outlines and summaries, and review questions.

58 Culver, Roger B.
ASTRONOMY
New York: Barnes & Noble, 1979, ISBN 0-06-460158-7 pbk. 295p; illus, diags. Glos (3), index (2).

This concise introductory text, part of the Barnes & Noble Outline Series, covers history, instrumentation, the principles of observational astronomy, motion and gravity, as well as the various features of the universe. Includes basic star maps.

59 Ebbighausen, E.G.
ASTRONOMY (5th ed.)
Columbus, OH: Merrill, 1985, ISBN 0-675-20413-5. 169p; illus, diags. Bibl (1), index (2).

A rather brief general text for the 'nonprofessional reader' using only basic algebra. The book's main goal is to make the student appreciate, 'how does the astronomer know?' Includes questions, and star charts.

*60 Johnson, F.
VOYAGE INTO ASTRONOMY
Dubuque, IA: Kendall/Hunt, 1977, pbk.

61 Kraus, John
OUR COSMIC UNIVERSE
Powell, OH: Cygnus-Quasar, 1980. 283p; illus (some col.), diags. Bibl (2), index (2).

A general introduction to astronomy by a pioneer of radio astronomy. The style of presentation falls between that of the basic introduction and the undergraduate text. The book is, according to the foreword by Arno Penzias, 'anecdotal, with an enjoyable series of human-interest glimpses showing how flashes of insight gained through the creation of new instruments have enlarged mankind's perception of the electromagnetic spectrum'.

62 Mehlin, Theodore G. and Charles A. Schweighauser
ASTRONOMY AND THE ORIGIN OF THE EARTH (3rd ed.)
Dubuque, IA: W. Brown, 1979, ISBN 0-697-05018-1 pbk. 150p; illus, diags. Index (2).

Intended for an introductory level course, this book concentrates on the Solar System. Includes chapter summaries. (2nd ed. 1973 only seen).

63 Moché, Dinah L.
ASTRONOMY: A SELF TEACHING GUIDE (2nd ed.)
New York: Wiley, 1981, ISBN 0-471-09713-6 pbk. 284 + 4p; illus, diags. Bibl (1), index (2).

A basic study course for the beginner, to be used alone or in conjunction with a textbook (provides cross references from chapters in this book to those in 16 popular US texts). Covers those topics most often found in introductory courses. The chapters are divided into short sections, each ending with a question for the reader, with further questions at the end of each chapter. Includes four detachable star maps.

64 Moore, Patrick
ASTRONOMY FOR O LEVEL (3rd ed.)
London: Duckworth, 1979, ISBN 0-7156-0732-4; 0-7156-0731-6 pbk. 216p; diags. Bibl (x), index (2).

An introduction for beginning students, though not a complete course. Includes general coverage of positional astronomy and planets and stars, with chapters on the history of astronomy, and satellites and space probes. There are questions and answers, and three projects: making a small telescope, and observation in the Solar System, and of the stars.

65 Oriti, Ronald A. and William B. Starbird
INTRODUCTION TO ASTRONOMY
Encino, CA: Glencoe Press, 1977, ISBN 0-02-478560-1. 402 + 23p; illus (some col.), diags. Glos (3), index (3).

An informal, non-mathematical text, which often uses a historical perspective, and includes a number of 'popular sidelights' (taking up almost 100 pages) on constellations and mythology, astrology, flying saucers, life in the universe, and the value of space exploration. Includes questions.

66 Pananides, Nicholas A. and Thomas Arny
INTRODUCTORY ASTRONOMY (2nd ed.)
Reading, MA: Addison-Wesley, 1979, ISBN 0-201-05674-7. 388 + 24p; illus (some col.), diags. Glos (3), index (2).

This text begins with two chapters on the history of astronomy and has a stronger than usual historical emphasis throughout. Includes chapter summaries and review questions.

67 Rogers, Eric M.
ASTRONOMY FOR THE ENQUIRING MIND: THE GROWTH AND USE OF THEORY IN SCIENCE
Princeton, NJ: Princeton University Press, 1982, ISBN 0-691-02370-0 pbk. 163p; diags. Bibl (x). Q

A slightly revised and corrected version of the astronomy section of the author's *Physics for the enquiring mind* (Princeton, 1960). Not intended to provide up to date information but 'to provide a clear example of the growth and use of *theory* in science'. Requires knowledge of elementary algebra. Includes problems, but no index.

68 Rosemergy, John C.
CELESTIAL HORIZONS: A CONCISE VIEW OF THE UNIVERSE
Boston: Allyn & Bacon, 1977, ISBN 0-205-05571-0 pbk. 260p; illus, diags. Bibl (x), glos (3), index (2).

This introductory text is deliberately short to provide students with a foundation upon which the instructor can build to suit a particular class. Selects only topics thought to be essential to an understanding of elementary astronomy. Takes a historical point of view, showing the development of our understanding of the universe, and outlining the establishment of physical principles. Includes review questions and suggested activities.

69 SCIENCE AND THE PLANET EARTH I (Science: a Foundation Course, Unit 1)
Milton Keynes: Open University, 1979, ISBN 0-335-08050-2 pbk. 46p; illus, diags. Q

This course text uses astronomy and the Earth to introduce the basic concepts of science, explaining the planet's shape, position in space, movement, and relationship to other bodies.

70 Swihart, Thomas L.
JOURNEY THROUGH THE UNIVERSE: AN INTRODUCTION TO ASTRONOMY
Boston: Houghton Mifflin, 1978, ISBN 0-395-25518-X. 366 + 16p; illus (some col.), diags. Bibl (1), glos (3), index (2). Q

Intended 'for anyone who is interested in astronomy'; the author deliberately sacrifices rigour and precision on occasion where clarity would otherwise be impeded. Not only non-mathematical, but even the use of numbers is kept to a minimum. Includes review and discussion questions, star maps.

71 Wagoner, Robert V. and Donald W. Goldsmith
COSMIC HORIZONS: UNDERSTANDING THE UNIVERSE
Stanford, CA: Stanford Alumni Association, 1982, ISBN 0-916318-08-8 pbk; San Francisco: Freeman, 1983, ISBN 0-7167-1417-5; 0-7167-1418-3 pbk. 195p; illus, diags. Bibl (2), index (2).

A non-mathematical text for the student and the general reader, which focuses on the wider universe and cosmology.

Standard/little or no mathematics used

72 Abell, George O.
DRAMA OF THE UNIVERSE
New York: Holt, 1978, ISBN 0-03-022401-2. 514 + 24 + 12p; illus (some col.), diags. Glos (3), bibl (2), index (3). Q

This text is presented in the form of a six-act play, with a prologue and epilogue, in which the material is arranged not topic by topic, but organized around concepts, as the following outline indicates: Act I: the development of Newtonian gravitational theory, II: the new physics of the 20th century, III: distances in space and measurement, IV: evolution of stars and the origin of the Solar System, V: the search for life, VI: the origin and evolution of the universe. The author incorporates some personal experiences within the text. Includes star charts.

73 Abell, George O.
EXPLORATION OF THE UNIVERSE (4th ed.)
Philadelphia: Saunders, 1982, ISBN 0-03-058502-3. 721 + 34 + 12p; illus (some col.), diags. Glos (3), bibl (3), index (3). Q

One of the most highly regarded textbooks of this type. It is specifically intended for a rather comprehensive course in astronomy for the liberal arts student, and uses some algebra. The author hopes it will be received as 'an introduction to the rational exploration of nature'. Includes exercises. The 21 appendices include data on stars and planets, and other supplementary information. There are six detachable star maps.

74 Abell, George O.
REALM OF THE UNIVERSE (3rd ed.)
Philadelphia: Saunders, 1984, ISBN 0-03-058504-X pbk. 560 + 24p; illus (some col.), diags. Bibl (2), glos (3), index (3). Q

A shorter version of (73). This edition has been completely revised and updated, with the order of topics changed slightly.

75 Apfel, Necia H. and J. Allen Hynek
ARCHITECTURE OF THE UNIVERSE
Menlo Park, CA: Benjamin/Cummins, 1979, ISBN 0-8053-4747-X pbk. 499 + 8p; illus (some col.), diags. Bibl (1), glos (3), index (2).

The authors hope to encourage 'the desirability of maintaining an open mind' by emphasizing what is not yet understood as well as what is known. After introducing the universe as a whole, and covering physical principles, the traditional arrangement is reversed, moving, in general, from great to small. Includes a section on UFOs. Discussion questions follow each chapter, with answers in an appendix. (This is a revised version of *Astronomy One*, 1972).

76 Bash, Frank N.
ASTRONOMY
New York: Harper & Row, 1977, ISBN 0-06-043853-3 pbk. 486 + 8p; illus (some col.), diags. Glos (2), index (2).

In this text, where the emphasis is on astrophysics, the author does not flinch from introducing basic mathematical astronomy. Begins with background physics and observational techniques, then moves through stellar astronomy to cosmology, and on to a discussion of the Sun and the Solar System. The text is broken up by numerous 'boxes', 'portfolios' and 'features' designed to explain or illuminate various phenomena and processes. Includes chapter summaries and exercises.

77 Berman, Louis and J.C. Evans
EXPLORING THE COSMOS (4th ed.)
Boston: Little, Brown, 1983, ISBN 0-316-09184-7 pbk. 543 + 16p; illus (some col.), diags. Glos (3), bibl (2), index (3). Q

The authors aim to give a solid background in a broad range of topics. Includes a section on exobiology. There are chapter summaries, review questions and star charts. A separate instructor's manual is available. (3rd ed., 1980 only seen).

78 Brandt, John C. and Stephen P. Maran
NEW HORIZONS IN ASTRONOMY (2nd ed.)
San Francisco: Freeman, 1979, ISBN 0-7167-1043-9. 614p; illus, diags. Bibl (1), index (3). Q

This text places 'emphasis on the relationship of life (and the individual) to the astronomical universe; the involvement of astronomy in fundamental questions that have occupied scientific philosophers, and the development of astronomical thought in the struggle for knowledge [against nature, man and his institutions]'. A separate 54-page study guide provides study objectives, key words, exercises, questions and assignments.

79 Chapman, Robert D.
DISCOVERING ASTRONOMY
San Francisco: Freeman, 1978, ISBN 0-7167-0034-4; 0-7167-0033-6 pbk. 518 + 8p; illus (some col.), diags. Glos (2), index (2). Q

The author attempts to avoid jargon for its own sake in this text, which features a strong historical approach. Includes review questions, but lacks a bibliography.

80 Dixon, Robert T.
DYNAMIC ASTRONOMY (4th ed.)
Englewood Cliffs, NJ: Prentice-Hall, 1984, ISBN 0-13-221333-8 pbk. 555 + 12p; illus (some col.), diags. Bibl (x), glos (3), index (2).

This text, aimed at a market wider than the student readership, is highly illustrated, and uses 'flip pages': several sequences of diagrams on the edges of the pages which give the illusion of movement when flipped through, to show, for example, the Ptolemaic planetary system, the retrograde motion of Jupiter, and an eclipsing binary star system; some are more effective than others. Includes questions, star charts.

81 Field, George B., Gerrit L. Verschuur and Cyril Ponnamperuma
COSMIC EVOLUTION: AN INTRODUCTION TO ASTRONOMY
Boston: Houghton Mifflin, 1978, ISBN 0-395-25321-7. 450 + 24p; illus

(some col.), diags. Glos (x), index (3). Q

Wider in scope than most introductory texts, this is virtually a textbook treatment of the ground covered in the Cosmic Evolution section (464–474). The authors' aim is to convey what is known about the evolution of the universe and its contents from the Big Bang to the emergence of intelligent life on Earth. Not arranged chronologically: chapters 1–8 give background astronomical information, chapter 9 covers the nature of space and time, chapter 10 deals with cosmology, with the next three chapters leading down to the origin of the Earth. The final three chapters look at the origin and evolution of life, extraterrestrial life, and the future. Requires some knowledge of algebra. Includes chapter summaries and 'thought questions', but lacks a bibliography.

82 Fredrick, Laurence W. and Robert H. Baker
AN INTRODUCTION TO ASTRONOMY (9th ed.)
New York: Van Nostrand, 1981, ISBN 0-442-22422-2. 502 + 8p; illus (some col.), diags. Bibl (x), glos (3), index (2). Q

This edition of a well established text has been extensively rewritten and rearranged, with expanded treatment of the planets, the Sun, and most other areas. Important terms appear in bold type. Includes questions. A separate instructor's manual (ISBN 0-442-22421-4) contains answers to the text questions, additional questions, comments, and tests.

83 Friedlander, Michael W.
ASTRONOMY: FROM STONEHENGE TO QUASARS
Englewood Cliffs, NJ: Prentice-Hall, 1985, ISBN 0-13-049867-X. 589 + 16p; illus (some col.), diags. Glos (3), bibl (3), index (3). Q

This text begins with several chapters which introduce the basics of astronomy from a historical perspective. The small mathematical content is mostly confined to separate boxes in the text. Includes chapter reviews, lists of new terms, problems; and supplementary data in the appendices.

84 Goldsmith, Donald
THE EVOLVING UNIVERSE (2nd ed.)
Menlo Park, CA: Benjamin/Cummins, 1985, ISBN 0-8053-3331-2. 562p; illus (some col.), diags. Bibl (x), glos (3), index (2).

This text reverses the traditional arrangement, and introduces the relevant physics as appropriate. Requires elementary algebra. Includes chapter summaries, lists of key terms, questions and projects. (This is a successor to the author's *The universe*, 1976). (1st ed. 1981 only seen).

*85 Hartmann, William K.
ASTRONOMY: THE COSMIC JOURNEY (3rd ed.)
Belmont, CA: Wadsworth, 1985, ISBN 0-534-04011-X. 553p; illus (some col.), diags. Bibl (3), index.

This virtually non-mathematical text presents the various bits of astronomical evidence, explains how they were observed or measured, and then fits them together to produce the grand picture. Presents only five equations, and includes problems based on them in appropriate chapters; they are: the small-angle equation, calculating circular and escape velocities, Wien's law, the Doppler effect, and the Stefan-Boltzmann law. (From review of 1st ed. *S&T* Oct. 1978, p 338–42).

*86 Hesse, Walter H.
OUR EVOLVING UNIVERSE
Belmont, CA: Dickenson, 1977, ISBN 0-8221-0184-X pbk. 481p; illus. Bibl, index.

87 Hodge, Paul W.
CONCEPTS OF CONTEMPORARY ASTRONOMY (2nd ed.)
New York: McGraw-Hill, 1979, ISBN 0-07-029147-0 pbk. 517 + 6p; illus (some col.), diags. Bibl (1), index (2). Q

An 'issue oriented' text, 'built around three vital astronomical questions: how did the Solar System form? How do stars evolve? and What is the nature of the universe at large?' Much background material has been moved to the 16 appendices. Good explanatory diagrams. Includes questions.

88 Jastrow, Robert and Malcolm H. Thompson
ASTRONOMY: FUNDAMENTALS AND FRONTIERS (4th ed.)
New York: Wiley, 1984, ISBN 0-471-89700-0. 564 + 24p; illus (some col.), diags. Glos (3), index (2). Q

This text begins with a discussion of stars, so that astrophysical knowledge can be brought to bear later when studying the Solar System. The fourth edition adds a chapter on the evolution of astronomical thought. Uses elementary algebra. Includes questions, and a 25-page appendix on methods of measurement.

89 Jeffreys, William H. and Robert R. Robbins
DISCOVERING ASTRONOMY
New York: Wiley, 1981, ISBN 0-471-44125-2. 466p; illus, diags. Index (2). Q

In this text the authors place far greater emphasis than others on practical work. Includes a set of simple astronomical instruments in an envelope inside the back cover. Requires only basic algebra. Activities are suggested wherever possible, mostly in the more traditional areas, and not relying on the interpretation of data gathered elsewhere. Includes questions (with selected answers), study objectives, chapter reviews, monthly star charts, and sample answer sheets for experiments.

90 Kaufmann, William J., III
ASTRONOMY: THE STRUCTURE OF THE UNIVERSE
New York: Macmillan, 1977, ISBN 0-02-362130-3. 491 + 4 + 13p; illus (some col.), diags. Glos (3), index (3).

'The purpose of this book is to present a clear, vivid non-mathematical treatment of the frontiers of modern astronomy'. Contains very brief coverage of the Solar System, which the author deals with in a separate book (608). Includes a chapter on history, but the strength of this text lies in its coverage of astrophysics, relativity and cosmology. There are questions and exercises, and pull-out monthly star charts.

91 Kaufmann, William J., III
UNIVERSE
New York: Freeman, 1985, ISBN 0-7167-1673-9. 594p; illus (some col.), diags. Bibl (x), glos (3), index (3). Q

In this well illustrated, non-mathematical text, the author's major objective is the student's ease of understanding. The traditional arrangement of subject matter can be divided into two parts: 1. early theories of the universe, and the planets, 2. stars, galaxies and cosmology. Unlike most textbooks the colour illustrations here are found throughout the text. Includes chapter summaries, questions (and answers), and essays by six astronomers. Supplementary information is presented in boxes. Instructor's guide: (*see* 18).

92 Long, Charles E.
DISCOVERING THE UNIVERSE
San Francisco: Harper & Row, 1980, ISBN 0-06-044034-1 pbk. 511 + 16p; illus (some col.), diags. Bibl (x), glos (2), index (2).

This text, which makes a greater effort than most to appeal to the general reader, is structured as 'a series of problems — puzzling observations with which astronomers have struggled'. The traditional arrangement has a strong historical emphasis. Includes chapter summaries, exercises, defined words in bold type, and puzzles (with solutions).

*93 Parker, Barry R.
CONCEPTS OF THE COSMOS: AN INTRODUCTION TO ASTRONOMY
San Diego, CA: Harcourt, Brace, Jovanovich, 1984, ISBN 0-15-512850-7. 516 + 16p; illus (some col.), diags. Bibl (x), index.

A non-mathematical text by a theoretical physicist who aims to present 'a clear, simple and interesting introduction' to astronomy. Relies on logical explanations of physical concepts, which leaves room for the lecturer to develop details and mathematics. Considerable attention is given to observational topics; seasonal star charts are provided. Includes chapter summaries, review and discussion questions. (From *S&T* Oct. 1984, p 325–6).

94 Pasachoff, Jay M.
ASTRONOMY: FROM THE EARTH TO THE UNIVERSE (2nd ed.)
Philadelphia: Saunders, 1983, ISBN 0-03-058419-1 pbk. 534 + 32p; illus (some col.), diags. Glos (3), bibl (2), index (3). Q

One of the better regarded texts, which began as a shorter version of (96) and (106). Heavily illustrated, and with numerous asides, serious and otherwise, to supplement and enliven the text. Emphasizes deep understanding of concepts and the process of astronomy. Stresses accuracy: there is a seemingly endless list of reviewers of this and the other versions, and of individuals who have made comments or suggestions. Very good colour illustrations.

95 Pasachoff, Jay M.
ASTRONOMY NOW
Philadelphia: Saunders, 1978, ISBN 0-7216-7100-4 pbk. 450 + 16p; illus (some col.), diags. Glos (3), bibl (2), index (3). Q

Another version of (96) and (106), this text begins with the stars and has less material on the history of astronomy and naked-eye observation.

96 Pasachoff, Jay M.
CONTEMPORARY ASTRONOMY (3rd ed.)
Philadelphia: Saunders, 1985, ISBN 0-03-071641-1 pbk. 530 + 20p; illus (some col.), diags. Bibl (2), glos (3), index (3). Q

The first edition of this work provided the basic text from which the others (94, 95 and 106) were developed. This version reverses the traditional arrangement. The third edition is nearly 100 pages shorter than the second, having been streamlined and reworked. But the number of questions and problems has been increased by 20 per cent. Mathematical sections are singled out for use at the teacher's discretion.

97 Protheroe, W.M., E.R. Capriotti and G.H. Newsom
EXPLORING THE UNIVERSE (3rd ed.)
Columbus, OH: Merrill, 1984, ISBN 0-675-20145-4. 586 + 32p; illus (some col.), diags. Glos (3), bibl (2), index (2).

This text emphasizes the evolution and continuing development of our understanding of the universe. Aims to show the process of science at work in the field of astronomy. A short story showing the human aspects of astronomy is used to introduce each chapter. Includes questions.

98 Seeds, Michael A.
HORIZONS: EXPLORING THE UNIVERSE (2nd ed.)
Belmont, CA: Wadsworth, 1985, ISBN 0-534-04017-9 pbk. 484 + 16 + 12p; illus (some col.), diags. Bibl (x), glos (2), index (2). Q

This text has a non-standard arrangement: a general introduction is followed by a discussion of the Earth-Moon-Sun system, history (in a separately numbered supplement), the tools of the astronomer, stars, galaxies, and cosmology, the origin of the Solar System, ending with a discussion on the possibility of life on other worlds. Non-mathematical text, but with some mathematical discussion in boxes. Makes good use of explanatory diagrams. New terms are printed in bold and listed after each chapter. Includes chapter summaries, questions, problems, and detachable monthly star charts. A software package and an instructor's manual are available separately.

99 Shipman, Harry L.
THE RESTLESS UNIVERSE: AN INTRODUCTION TO ASTRONOMY
Boston: Houghton Mifflin, 1978, ISBN 0-395-25392-6. 464 + 10p; illus (some col.), diags. Bibl (x), glos (2), index (2). Q

The author focuses this text on contemporary issues to bring the concepts mentioned in the media closer to the student. Uses minimal mathematics, though the applications of numerical data are shown in separate boxes. Emphasizes stellar evolution in favour of stellar properties. Includes chapter summaries, key concepts and review questions, and, in the appendices, activities for amateurs, a crossword puzzle, and a guide to the pronunciation of selected astronomical names.

100 Snow, Theodore P.
THE DYNAMIC UNIVERSE: AN INTRODUCTION TO ASTRONOMY (2nd ed.)
St. Paul, MN: West, 1985, ISBN 0-314-88512-9. 601 + 64p; illus (some

col.), diags. Bibl (x), glos (3), index (2). Q

This introductory text includes many supporting features to the basic text. Besides the many eye-catching colour photographs there are seven 'guest editorials' by astronomers (six of which have been replaced by new essays by new authors in the second edition), numerous 'astronomical insights' spread throughout the book, which provide historical or other background information, chapter summaries, lists of topics, perspective comments, and review questions. Also available are a study guide by Catharine D. Garmany and the author, and an instructor's manual by Stephen J. Shawl.

101 Snow, Theodore P.
ESSENTIALS OF THE DYNAMIC UNIVERSE: AN INTRODUCTION TO ASTRONOMY
St. Paul, MN: West, 1984, ISBN 0-314-77798-9 pbk. 489 + 24p; illus (some col.), diags. Bibl (x), glos (3), index (2). Q

A shorter version of (100) whose chief exclusion is the discussion of the history of astronomy.

102 Waxman, Gerald D.
INTRODUCTORY ASTRONOMY
Geneva, IL: Paladin House, 1978, ISBN 0-88252-078-4. 628p; illus, diags. Glos(3). Q

This introductory text has a strong emphasis on practical work; it includes many classroom exercises. The equipment needed may be constructed from plans given in the appendices. There is no index.

103 Zeilik, Michael
ASTRONOMY: THE EVOLVING UNIVERSE (4th ed.)
New York: Harper & Row, 1985, ISBN 0-06-047374-6. 494 + 8p; illus (some col.), diags. Bibl (x), glos (3), index (3). Q

This highly regarded text is divided into four parts: 1. changing conceptions of the cosmos, 2. the planets past and present, 3. the universe of stars and galaxies, 4. cosmic evolution. Lists learning objectives at the beginning of each chapter, key concepts at the end. Includes study exercises. A separate instructor's manual (with Richard Reif) is available.

Physical principles emphasized/mathematics used

104 Hoyle, Fred and Jayant Narlikar
THE PHYSICS-ASTRONOMY FRONTIER
San Francisco: Freeman, 1980, ISBN 0-7167-1160-5. 483 + 8p; illus

(some col.), diags. Glos (2), index (2).

This text, which uses little mathematics, deals with astronomy from the point of view of the physicist, stressing topics such as the nature of atoms, quantum mechanics and radiation. According to the authors the book 'should easily be within the scope of anyone who has a strong high school background in general science'. Does not cover the Solar System. Includes general problems and questions.

105 Motz, Lloyd and Anneta Duveen
ESSENTIALS OF ASTRONOMY (2nd ed.)
New York: Columbia University Press, 1977, ISBN 0-231-04009-1. 763p; illus, diags. Bibl (2), indexes (3). Q

A comprehensive text which requires knowledge of at least elementary algebra, but is less demanding than (109). Makes great use of explanatory diagrams. This is not a fully revised version of the first edition of 1966; there is a separate new section (part 5) for developments since 1966, with section headings corresponding to those in the main text, where revisions have been kept to a minimum.

106 Pasachoff, Jay M. and Marc L. Kutner
UNIVERSITY ASTRONOMY
Philadelphia: Saunders, 1978. ISBN 0-7216-7099-7. 764 + 44 + 27 + 24p; illus (some col.), diags. Glos (3), bibl (2), index (3). Q

An expanded version of (96, 1st ed.), particularly the coverage of the Earth, relativity, and basic equations, requiring knowledge of algebra but not calculus. Background discussions of physics and astronomy appear as needed, rather than as introductory chapters. Stresses understanding, in several cases describing the actual work of an astronomer. The book begins with a discussion of the stars, the Sun and stellar evolution, then covers the Solar System, before moving outward to deal with interstellar matter and extragalactic astronomy. Includes questions.

107 Roy, A.E. and D. Clarke
ASTRONOMY: PRINCIPLES AND PRACTICE (2nd ed.)
Bristol: Adam Hilger, 1982, ISBN 0-85274-463-3; 0-85274-464-1 pbk. 342p; illus, diags. Bibl (1), index (2). Q

First of a two-volume set which aims 'to fulfil the need of a preliminary science course or a liberal arts course at the first year university level'. This volume prepares the student for its companion (108) by presenting 'the basic software and hardware of the subject'. Contains mathematical treatment at a level which many US texts avoid, but is not so difficult as to deter the less mathematically inclined. Includes exercises and practical work, with problems in the text, many of which are worked out in detail. New terms are shown in bold type. Makes good use of explanatory diagrams.

108 Roy, A.E. and D. Clarke
ASTRONOMY: STRUCTURE OF THE UNIVERSE (2nd ed.)
Bristol: Adam Hilger, 1982, ISBN 0-85274-465-X; 0-85274-466-8 pbk. 296p; illus, diags. Bibl (1), index (2). Q

Second of a two-volume set which gives 'a description of the universe as revealed by the application of the methods discussed in the first book' (107). This volume is divided into three parts; the first covers the Solar System, the second describes the constituents of the Galaxy: stars and interstellar material, the third deals with our galaxy and other galaxies, ending with a discussion of cosmological ideas. Includes problems in the text (with answers).

109 Shu, Frank H.
THE PHYSICAL UNIVERSE: AN INTRODUCTION TO ASTRONOMY
Mill Valley, CA: University Science Books, 1982, ISBN 0-19-855706-X. 584 + 16p; illus (some col.), diags. Index (3). Q

This text, based on lectures from an introductory astronomy course, is one of the most rigorous and demanding of such works. The author relates astronomical facts firmly to fundamental physical principles. The arrangement is non-standard, with an attempt to give equal emphasis to each area covered, namely, basic principles, stars, galaxies and cosmology, the Solar System, and life. The book is intended for use on several different levels of intellectual attainment, so includes problems of widely varying difficulty.

110 Unsöld, Albrecht and Bodo Baschek, (R.C. Smith, trans., based on trans. of 1st ed. by W.H. McCrea)
THE NEW COSMOS (3rd ed.)
New York: Springer-Verlag, 1983, ISBN 0-387-90886-2 pbk (trans. of 3rd German ed. 1983). 479p; illus, diags. Bibl (3), index (3).

Aims 'to give the student of astronomy, the scientist who is not a specialist, and the amateur astronomer an accessible but up-to-date introduction to astronomy and astrophysics'. Requires knowledge of at least elementary algebra. The arrangement of topics is basically standard, moving from classical astronomy and the Solar System, through astrophysics, to cosmogony and cosmology. The final part is a summary of selected recent results, covering the Solar System, spectroscopy in the extreme ultra-vioiet, x-ray and gamma-ray astronomy. The bibliography contains no recent works.

111 Wyatt, Stanley P.
PRINCIPLES OF ASTRONOMY (3rd ed.)
Boston: Allyn & Bacon, 1977, ISBN 0-205-05679-2. 730 + 20p; illus (some col.), diags. Glos (3), bibl (2), index (2). Q

This text, designed for a comprehensive full-year course, assumes knowledge of algebra and geometry, but introduces and explains integral logarithms. A solid, reliable work, which rejects the over-simplified, sometimes gimmicky approach of other textbooks.

112 Wyatt, Stanley P. and James B. Kaler
PRINCIPLES OF ASTRONOMY: A SHORT VERSION (2nd ed.)
Boston: Allyn & Bacon, 1981, ISBN 0-205-07315-8. 536 + 16p; illus (some col.), diags. Glos (3), bibl (2), index (2). Q

Shorter, but no less rigorous than (111). The bibliography concentrates on presenting more advanced readings. Includes problems. A 20-page instructor's manual (0-205-07316-6 pbk) is available.

*113 Zeilik, Michael and John Gaustad
ASTRONOMY: THE COSMIC PERSPECTIVE
New York: Harper & Row, 1983, ISBN 0-06-047387-8. 954 + 64 + 8p; illus (some col.), diags. Glos (3), bibl, index.

An expanded and rearranged version of (103, 3rd ed.), whose level is increased by the use of mathematics for more rigorous explanations of astrophysical concepts. It is intended for 'science-oriented students, though not necessarily science majors'. The book's five sections cover: 1. the sky visible to the naked eye and the history of astronomy, 2. stars and stellar evolution, 3. the Solar System, 4. the larger structure of the universe and cosmology, and 5. 'cosmic speculation' (life in the universe, and the future of the universe, our galaxy, the Solar System, and the human race). (From *S&T* Jan. 1984, p 38–9).

Miscellanies

114 Bath, Geoffrey (ed.)
THE STATE OF THE UNIVERSE
Oxford: Oxford University Press, 1980, ISBN 0-19-857549-1. 199 + 8p; illus (some col.), diags. Bibl (x), index (1).

A collection of articles arising from a series of lectures held at

Wolfson College, Oxford, in 1979. Requires some knowledge of algebra. The articles cover the origin of the universe, galaxies, stars, the origin of the elements, the x-ray universe, black holes, planetary exploration, and the new astronomies.

115 Brandt, John C. and Stephen P. Maran (ed.)
THE NEW ASTRONOMY AND SPACE SCIENCE READER
San Francisco: Freeman, 1977, ISBN 0-7167-0350-5; 0-7167-0349-1 pbk. 371p; illus, diags. Bibl (2), index (2). Q

A collection of 44 articles reprinted from various sources, intended to introduce students to current topics of astronomical research, and to supplement what may be sketchy coverage in formal textbooks. The level of the articles ranges from popular to fairly technical. Topics covered include the history of astronomy, the Solar System, stars and stellar evolution, interstellar matter and the galaxy, galaxies and the universe, astronomy beyond the visible spectrum, and astronomy from space.

116 Campbell-Jones, Simon (ed.)
HORIZON: AT THE EDGE OF THE UNIVERSE (UK)/AT THE EDGE OF THE UNIVERSE (US)
London: BBC (Ariel Books), 1983, ISBN 0-563-17954-6 pbk; New York: Universe Books, 1983, ISBN 0-87663-433-1. 216p; illus, diags.

Nine articles based on programmes in the BBC TV *Horizon* series, written by some of its producers. Covers selected topics in a general movement from the Earth outwards. Each article focuses on the scientists, technologists and astronauts involved, in an attempt to show the process of discovery at work.

117 Corliss, William R. (compiler)
MYSTERIOUS UNIVERSE: A HANDBOOK OF ASTRONOMICAL ANOMALIES
Glen Arm, MD: The Sourcebook Project, 1979, ISBN 0-915554-05-4. 710p; diags. Index (2).

A compendium of descriptions of unusual astronomical phenomena, compiled from hundreds of issues of astronomy and general science journals, popular as well as professional, so the sources may not always be reliable. Not for continuous reading but interesting to dip into. A further compilation of natural curiosities by the author is *Rare halos, mirages, anomalous rainbows: a catalog of geophysical anomalies* (The Sourcebook Project, 1984, ISBN 0-915554-12-7, 238p).

118 Cornell, James and Alan P. Lightman (ed.)
REVEALING THE UNIVERSE: PREDICTION AND PROOF IN ASTRONOMY
Cambridge, MA: MIT Press, 1982, ISBN 0-262-03080-2; ISBN 0-262-53043-0 pbk. 246p; illus, diags. Bibl (x), index (2).

An unusual book which arose from a series of public lectures given at the Boston Museum of Science in 1980. Explores 'the relationship between theory and observation within the context of several topics of current interest in astronomy and astrophysics'. Each of the six topics is tackled by two scientists, one predominantly a theorist, the other an experimentalist. Covers Einstein's perception of space and time, the evolution of the Solar System, the Sun's corona, quasars and black holes, x-ray burst sources, and the age and structure of the universe. Interesting insights into the development of scientific knowledge.

119 Henbest, Nigel
THE MYSTERIOUS UNIVERSE (UK)/MYSTERIES OF THE UNIVERSE (US)
London: Ebury Press, 1981, ISBN 0-85223-212-8; New York: Van Nostrand, 1981, ISBN 0-442-23690-5; 1983, ISBN 0-442-23636-0 pbk. 184p; illus (some col.), diags. Glos (1), index (2). Q

Covers the whole field of astronomy, but is not a standard popular descriptive account; it 'is a round up of some mysterious aspects of the universe as [the author] sees them'. Selective, but wide ranging, the topics covered include cosmic influences on the Earth, the possibility of extraterrestrial life, the Sun's lack of neutrinos, and quasars.

120 Henbest, Nigel (ed.)
OBSERVING THE UNIVERSE
Oxford: Blackwell, 1984, ISBN 0-85520-727-2; 0-85520-726-4 pbk. 288p; illus, diags. Index (2).

A collection of 42 articles and news items from *New Scientist*, covering observational aspects of astronomy (as carried out by astronomers) from its ancient beginnings to the present investigations in non-visible wavelengths of light, with emphasis on the technological aspects, including the use of computers.

121 Maffei, Paolo, (Mirella and Riccardo Giacconi, trans.)
MONSTERS IN THE SKY
Cambridge, MA: MIT Press, 1980, ISBN 0-262-13153-6; New York: Avon, 1981, ISBN 0-380-55517-4 pbk (trans. of 2nd Italian ed. 1977). 342p; illus, diags. Bibl (2), index (3).

A sequel to (47), which, despite its ludicrous title, is a serious study of cosmic anomalies intended for the general reader with some background knowledge of astronomy. The topics covered include comets, tektites, the Eta Carinae nebula, supernovae, black holes and white holes, BL Lacertae objects, Seyfert galaxies, and quasars.

122 Man, John (general ed.)
THE ENCYCLOPEDIA OF SPACE TRAVEL AND ASTRONOMY
London: Octopus, 1979, ISBN 0-7064-0992-2. 224p; illus & diags (mostly col.). Glos/index (2). Q

This profusely illustrated coffee-table book contains twelve articles by well known astronomy popularizers, such as Ian Ridpath, Nigel Henbest and John Gribbin. The emphasis is on astronomy, not space travel as the title suggests. A possible introduction for the complete beginner.

123 Messel, H. and S.T. Butler (ed.)
FOCUS ON THE STARS
London: Heinemann, 1977, ISBN 0-435-68282-2; New York: Crane, Russak, 1978, ISBN 0-8448-1299-4. 291p; illus, diags. Index (2).

A collection of articles covering various aspects of astronomy and space science, based on a series of lectures for high school students given at the University of Sydney. They are rather older than the date of the book suggests.

124 Moore, Patrick (ed.)
MODERN ASTRONOMY: SELECTIONS FROM THE YEARBOOK OF ASTRONOMY
London: Sidgwick & Jackson, 1977, ISBN 0-283-98322-1; 0-283-98369-8 pbk; New York: Norton, 1977, ISBN 0-393-06417. 184p; illus, diags.

Eighteen articles from the *Yearbook* (214), not intended for beginners, covering history and instruments, the Solar System, stars, the Galaxy, and extra-galactic astronomy. They are, in effect, reviews of selected aspects of astronomical research for the amateur.

125 Moore, Patrick
PATRICK MOORE'S ARMCHAIR ASTRONOMY
Wellingborough: Patrick Stephens, 1984, ISBN 0-85059-718-8; New York: Norton, 1985, ISBN 0-393-02253-6. 185p; illus, diags. Glos (2), index (2).

A collection of 126 entertaining 'snippets', touching on all aspects of astronomy and space exploration, with a strong historical content.

126 Moore, Patrick
THE SKY AT NIGHT 6
London: BBC, 1978, ISBN 0-563-17432-3. 192p; illus, diags. Index (1).
A collection of 34 short pieces, chronologically arranged, based on TV programmes transmitted by the BBC from January 1975 to September 1977. Covers a wide variety of topics. Previous volumes in the series appeared in 1964, 1968, 1970, 1972 and 1976.

127 Moore, Patrick
THE SKY AT NIGHT 7
London: BBC, 1980, ISBN 0-563-17860-4 pbk. 210 + 16p; illus, diags. Index (1).
This volume contains 32 articles based on programmes transmitted by the BBC between September 1977 and March 1980.

128 Moore, Patrick
THE SKY AT NIGHT [8]
Wellingborough: Patrick Stephens, 1985, ISBN 0-85059-753-6 pbk. 192p; illus. Index (2).
Contains 50 articles based on programmes transmitted by the BBC between April 1980 and November 1984.

129 Mumford, George S.
THE CLOUDY NIGHT BOOK
Cambridge, MA: Sky, 1979, ISBN 0-933346-00-X pbk. 115p; illus.
For nights when the weather turns the telescope into an ornament, this collection of games, puzzles, 'cloudy night lore', cartoons, acrostics and quizzes will keep the keen amateur entertained and, the author hopes, may also provide something of educational value.

130 Seeds, Michael A. (ed.)
ASTRONOMY: SELECTED READINGS
Menlo Park, CA: Benjamin/Cummins, 1980, ISBN 0-8053-8531-3 pbk. 178p; illus & diags (mostly col.). Bibl (2), index (2). Q
A profusely illustrated collection of articles reprinted from *Astronomy* (135), correlated to the chapters of 27 textbooks. Primarily for students, to 'reinforce and sustain' their interest.

131 Skinner, Brian J. (ed.)
THE SOLAR SYSTEM AND ITS STRANGE OBJECTS:

READINGS FROM *AMERICAN SCIENTIST*
Los Altos, CA: William Kaufmann, 1981, ISBN 0-913232-84-X pbk. 193p; illus & diags (some col.). Bibl (x), index (1). Q

A peculiarly titled collection of 19 articles published 1959–1980 (mostly late 1970s), covering various aspects of astronomy, astrophysics and space science: 'authoritative statements by specialists'. Not for the casual reader.

132 Verschuur, Gerrit L.
COSMIC CATASTROPHES
Reading, MA: Addison-Wesley, 1978, ISBN 0-201-08098-2; 0-201-08099-0 pbk. 214p; illus. Index (2).

The author considers seven types of catastrophe, such as a supernova, a meteor impact, and a black hole, each in two chapters: the first is a short science fiction story, the second provides the scientific rationale.

133 Verschuur, Gerrit L.
STARSCAPES: TOPICS IN ASTRONOMY
Boston: Little, Brown, 1977, pbk. 202p; illus, diags. Glos (2).

A collection of 27 brief essays on areas of 'current' interest, mostly from *Astronomy* (135). Concentrates on knowledge gained from astronomy beyond the visible spectrum.

134 West, Richard M. (ed.)
UNDERSTANDING THE UNIVERSE: THE IMPACT OF SPACE ASTRONOMY
Dordrecht: Reidel, 1983, ISBN 90-277-1647-1. 249p; illus, diags. Bibl (x), index (2).

A collection of five articles based on lectures given at UNISPACE 82 (*see* 751), reproduced from typescript. Intended to present an account of advances in understanding the universe resulting from a combination of space-borne and ground-based astronomy. Aimed at the 'interested layman', but is fairly technical in places. Covers 'the atmosphere of the Sun and the stars' (text in French, notes to figures in English), 'solar-terrestrial influence', 'the size, shape and temperature of the stars', 'x-ray and gamma-ray astronomy', and 'space science and cosmology'.

Periodicals

135 ASTRONOMY
(AstroMedia Corp., 625 E. St. Paul Avenue, PO Box 92788, Milwaukee, WI 53202, USA), 1973–, monthly, ISSN 0091–6358. 90p; illus & diags (mostly col.). Ads. Q

The subtitle on the title page, 'the world's most beautiful astronomy magazine', indicates its emphasis on presenting attractive colour illustrations: it is clearly designed to appeal to the general reader, as well as the less advanced amateur. Each issue contains a number of readable articles and features similar to those in *Sky & Telescope* (143): news of research, information for observers, a monthly star chart, reviews of books and other items, and a free small advertisements service. AstroMedia also publish *Odyssey*, a magazine for younger readers. (*See also* 130 and 133).

136 THE ASTRONOMY QUARTERLY
(Circulation Dept., Astronomy Quarterly, PO Box 35549, Tucson, AZ 85740, USA), 1977–, 2 issues/year, ISSN 0364-9229. Diags. Some ads.

A scholarly journal for the 'educated layman and student', which 'examines the role of astronomy, especially cosmology, in natural philosophy and in culture in general as well as astronomy's influence upon the intellectual atmosphere of the epoch'. Despite the title it is published only twice a year. Includes book reviews. An index appears in the last number of each volume of four issues.

137 GRIFFITH OBSERVER
(2800 East Observatory Road, Los Angeles, CA 90027, USA), 1937–, monthly, ISSN 0195-3982. Illus, diags.

A small, popular-level magazine featuring well illustrated, usually brief, articles. But also some longer ones: the July 1985 issue features a single long article on Russian attempts to beat the US to the Moon. Occasional special issues, e.g. on archaeoastronomy (June 1983). Includes book reviews in some issues, and monthly basic sky maps.

138 McDONALD OBSERVATORY NEWS
(RLM 15.308, The University of Texas at Austin, Austin, TX 78712, USA), 1973–, monthly, ISSN 0745-1997. 8p; illus & diags (some col.). Q

'A popular level newsletter on astronomy'. Includes a feature article; notes for naked-eye observers, with monthly star maps; MERLIN, a readers' question and answer column; and hints on using a small telescope.

139 MERCURY
(Astronomical Society of the Pacific, 1290 24th St., San Francisco, CA 94122, USA), 1972–, bi-monthly. Illus, diags. Some ads. Q

The journal of the ASP is 'a non-technical journal intended for the general reader, amateur astronomer and science educator'. It 'features articles on current developments in astronomy research, connections between astronomy and other fields, innovations in astronomy education, . . . interviews with astronomers', history and Society matters. Features bibliographic items, including an annual round up of the previous year's books. Often publishes articles debunking pseudoscience. Unfortunately not widely available in the UK.

An index to astronomy articles in *Scientific American* (1960–1981) appeared in March/April 1981, p 58–63. (*See also* 22).

140 POPULAR ASTRONOMY
(Martin Ratcliffe, Secretary, Junior Astronomical Society, 10 Swanwick Walk, Tadley, Basingstoke, Hants, RG26 2JZ, UK, *or* Guy Fennimore, 42 Dunster Road, West Bridgford, Nottingham, NG2 6JE, UK), 1981–, quarterly. Illus, diags. Ads. Q

Originally titled *Hermes* (1953–80), the journal of the JAS (not only for younger members, but, in general, for less advanced amateurs of all ages) contains non-technical articles on all aspects of astronomy, including observing and education. Features include book reviews and letters. Members also receive six issues of the *News Circular* each year, with extra issues for important news. The JAS has a number of sections: Aurora, Comet, Deep Sky, Lunar (which publishes a six-monthly newsletter, *Luna*), Meteor, Occultation, Photographic, Planetary, Satellite, Solar, and Variable Star.

141 QUARTERLY JOURNAL OF THE ROYAL ASTRONOMICAL SOCIETY
Oxford: Blackwell, 1960–, quarterly, ISSN 0035-8738 (RAS, Burlington House, Piccadilly, London, W1V 0NL, UK). Illus, diags.

'Publishes the domestic news of the Society as well as articles and conference reports suitable for the non-specialist. The review papers and conference reports are intended for a wide readership, and are written from a personal point of view'. Includes articles on peripheral issues such as education. Background astronomical knowledge is required.

142 THE REFLECTOR: NEWSLETTER OF THE ASTRONOMICAL LEAGUE
Loves Park, IL: Forest City Letter Service, 1952–, quarterly, ISSN 0034-2963 (Editorial address: Carol J. Beaman, 6804 Alvina Road, Rockford, IL 61103, USA). About 8p; illus, diags. Ads. Q

The Astronomical League is an association of US astronomy societies; this newsletter is sent to individual members of affiliated societies. Contains national and regional news of activities and individuals, League business matters, and book reviews. Since February 1984 a four-page insert, *Dark Skies for Comet Halley Journal*, has been included. Several other publications are available from the Astronomical League (*see* 17, 221, 241, 251 and 270).

143 SKY & TELESCOPE
(Sky Publishing Corp., 49 Bay State Road, Cambridge, MA 02238-1290, USA), 1941–, monthly, ISSN 0037-6604. Illus & diags (some col.). Ads. Q

The world's best known and, generally speaking, best loved astronomy magazine. It is aimed mainly at the amateur and the student. In over 100 pages it contains articles on general astronomical, and sometimes space science topics; news of research, news for amateurs, observing information, a feature on telescope making and observatories for the amateur, a bi-monthly feature for beginners, and covers astronomical computing, and history. Includes correspondence and very good book reviews. There is a monthly pull-out star chart. The many advertisements feature mostly equipment.

Many articles published before 1977 were reprinted in the nine-volume '*Sky & Telescope* Library of Astronomy', edited by Thornton Page and Lou Williams Page (New York: Macmillan, 1965–69, 1976). Two reprints of material from *S&T* are: Jean Meeus, *Some bright visual binary stars*; and Charles E. Worley, *Visual observing of double stars*. (*See also* 226, 259, 336 and 364).

Other Works

General

144 Brown, Hanbury
MAN AND THE STARS
Oxford: Oxford University Press, 1978, ISBN 0-19-851001-2. 185p; illus, diags. Glos (2), bibl (2), indexes (2). Q

The author tries to answer the question, 'what use is astronomy?' in a book arising from his work as head of the Narrubri Observatory in New South Wales. Emphasizes the historical development of astronomy in showing how observation of the stars has helped and influenced the development of mankind, with respect to the calendar, time, navigation, science (from the laws of motion to relativity), and cosmology.

*145 Callender, John D.
OBSERVATORIES
Medina, WA: Russell-Winthrop, 1978, pbk. 64p.

Little more than a checklist of 21 major observatories around the world that an amateur might wish to visit, this booklet provides a rough identifying photograph and one or two paragraphs of description for each one. (From *S&T* June 1979, p 575).

A more satisfactory guide, but only for American establishments, is: H.T. Kirby-Smith, *US observatories: a directory and travel guide* (New York: Van Nostrand, 1976, ISBN 0-442-24451-7; 0-442-24450-9 pbk, 173p).

146 Cohen, Martin
IN QUEST OF TELESCOPES
Cambridge, MA: Sky, 1980, ISBN 0-933346-25-5; Cambridge: Cambridge University Press, 1982, ISBN 0-521-24989-9. 131p; illus, diags. Index (2).

An interesting behind the scenes look at astronomy: the author's 'astronomical autobiography, as far as 1979. It tells you what it is like to lead the life of a professional astronomer'. Begins with the development of his interest as a schoolboy in England, then via Cambridge and the Royal Greenwich Observatory, to a number of US observatories, including some of the biggest and most important in the world.

147 Hagar, Charles F.
PLANETARIUM: WINDOW TO THE UNIVERSE
Oberkochen, West Germany: Carl Zeiss, 1980, ISBN 3-87517-005-9. 193p; illus (mostly col.), diags. Bibl (2). Q

A glossy, lavishly illustrated book about planetaria, sponsored by Carl Zeiss, suppliers of planetarium projectors worldwide. Looks behind the scenes, covering engineering and technical aspects, control systems and special effects, the Zeiss planetarium, the shows, education, and the planetarium and the community. Includes a catalogue of Zeiss planetaria, with pictures and notes.

148 Harwit, Martin
COSMIC DISCOVERY: THE SEARCH, SCOPE AND HERITAGE OF ASTRONOMY
New York: Basic Books, 1981, ISBN 0-465-01428-3; Brighton: Harvester Press, 1981, ISBN 0-7108-0089-4; 0-7108-0625-6 pbk; Cambridge, MA: MIT Press, 1984, ISBN 0-262-58068-3 pbk. 334p; illus, diags. Bibl (3), glos/index (3).

An interesting and important survey of the process of astronomical discovery, intended for 'a wide range of readers interested in [this subject] from the astronomer's, the historian's, and the policymaker's points of view.' Begins with a survey of the author's findings in this study, followed by descriptions of extra-Solar objects and how they were discovered. He next looks at the role of improved technology in making discoveries, and considers the possible contribution of future innovations. Moving to the question of new phenomena, he asks how do we know they are new and not merely variations of known phenomena?, and discusses the limits of astronomical observation. The final chapter considers the best way of continuing future searches, technological gaps, and manpower.

149 Malin, David and Paul Murdin
COLOURS OF THE STARS
Cambridge: Cambridge University Press, 1984, ISBN 0-521-25714-X. 198p; illus & diags (mostly col.). Bibl (1), index (1). Q

A profusely illustrated book aimed at the serious amateur, as well as the general reader. Shows the importance of colour in the study of astronomy: how colours are seen and photographed, and how they enhance our understanding of stars, nebulae and galaxies. Describes the techniques used by professional astronomers.

150 Marx, Siegfried and Werner Pfau, (C.S.V. Salt, trans.)
OBSERVATORIES OF THE WORLD
New York: Van Nostrand, 1982, ISBN 0-442-26270-1; Poole: Blandford, 1982, ISBN 0-7137-1191-4. 200p; illus (some col.), diags. Bibl (2), index (1).

A descriptive guide to 40 selected observatories, covering history, the site, equipment, and the type of research carried out. Highly illustrated, so the text is fairly brief. In addition the illustrations often appear in blocks, not always next to the pertinent text. An introduction deals with types of telescope, and the methods of astronomical research. The bibliography includes a number of articles from *Sky & Telescope*, but also many which are rather esoteric. (This translation of the 1st German ed. 1979, was edited by Simon Mitton).

Picture books/art

151 Ferris, Timothy with Carolyn Zecca
SPACESHOTS: THE BEAUTY OF NATURE BEYOND EARTH
New York: Pantheon Books, 1984, ISBN 0-394-53890-0. 143p; illus (some col.). Bibl (1). Q

A fine collection of astronomical and space images. The explanatory captions are gathered at the back of the book.

152 Hapgood, Fred
SPACESHOTS: AN ALBUM OF THE UNIVERSE
New York: Times Books, 1979, ISBN 0-8129-0823-6 pbk. 79p; illus (mostly col.). Q

A general selection of astronomical images in the visible part of the spectrum, mostly full- or double-page, arranged from the Earth outward.

153 Iwasaki, Kazuaki and Isaac Asimov
VISIONS OF THE UNIVERSE
Montrose, CA: The Cosmos Store, 1981, ISBN 0-939540-01-0. 47p; col. illus. Q

A collection of astronomical paintings by Iwasaki, with captions by Asimov; the full-page illustrations are placed opposite the text. Covers most aspects of astronomy, as well as space exploration and the evolution of the Earth.

154 Miller, Ron
SPACE ART
New York: Starlog Press, 1978, ISBN 0-931064-06-6; 0-931064-04-X pbk. 192p; illus (mostly col.). Bibl (1). Q

A compendium of astronomical and space artwork by various artists from the 19th century to the present. Arranged by theme, from the Solar system outward. Includes brief captions, an index to the artists, and a section on the making of a space painting.

155 Miller, Ron and Frederick C. Durant, III
WORLDS BEYOND: THE ART OF CHESLEY BONESTELL
Norfolk, VA: Donning, 1983, ISBN 0-89865-225-1; 0-89865-195-6 pbk; 0-89865-226-1 limited ed. 133p; illus (mostly col.). Q

About half of this book is devoted to the astronomical art of this long-lived artist (1888–), who trained first as an architect. Includes a brief biography by Durant; the rest of the text is redundant in the shadow of Bonestell's almost photographic representations of worlds unseen by human eyes, some of which have turned out to be remarkably accurate.

156 SPACE IMAGES
New York: Lustrum Press, ISBN 0-912810-37-8; 0-912810-36-X pbk. 99p; illus (some col.).

A collection of images from JPL/NASA, first presented at the Grey Art Gallery and Study Centre, New York 1981. Includes a brief essay on 'The photography of space exploration'.

2 Practical Astronomy

Beginner's Guides

Basic

157 Boulton, John
BASIC STEPS IN ASTRONOMY
Poole: Blandford, 1979, ISBN 0-7137-1012-8. 144 + 4p; illus (some col.), diags. Index (1).

A rather brief guide to practical, as well as general, astronomy, all outlined inadequately in about 100 pages. Includes 24 pages of star maps.

158 Bullock, James Benbow
STARS FOR LINCOLN, DOCTORS AND DOGS
San Francisco: Gourmet Guides, 1981, ISBN 0-937024-00-7 pbk. 96p; diags.

The author, a sculptor and poet, guides the reader through the constellations with the aid of finder charts, while discoursing on their mythological background. Illustrated with cartoons.

159 Burnham, Robert
THE STAR BOOK
Milwaukee, WI: AstroMedia, 1983, ISBN 0-913135-00-3; Cambridge: Cambridge University Press, 1984, ISBN 0-521-25833-2, both spiral-bound. 17p; illus, diags. Q

Eight half-seasonal star charts for the Northern observer, on durable paper suitable for outdoor use, each facing a page of concise descriptive text covering notable features and constellation lore (with some finder maps).

*160 Chandler, David
EXPLORING THE NIGHT SKY WITH BINOCULARS
La Verne, CA: D. Chandler, 1984, ISBN 0-9613207-0-2 pbk. 45p; diags.

A companion to the author's *The night sky* planisphere (Cambridge, MA: Sky). A small but wide-ranging guide which takes apprehensive beginners by the hand and leads them through the highlights of the cosmic scene. (From *S&T* Feb. 1985, p 131).

*161 Claibourne, Robert
THE SUMMER STARGAZER: ASTRONOMY FOR BEGINNERS
New York: Penguin, 1981, ISBN 0-14-046487-5 pbk (reprint of 1st hardcover ed. 1975). 222p; diags. Bibl (2), index.

*162 Cleminshaw, Clarence H.
THE BEGINNER'S GUIDE TO THE SKIES: A MONTH-BY-MONTH HANDBOOK FOR STARGAZERS AND PLANET WATCHERS
New York: Crowell, 1977, ISBN 0-690-01214-4. 152p; illus. Index.

*163 Couper, Heather and Ronald Alpiar
STAR FINDER
London: Century, 1984, ISBN 0-7126-0571-1 pbk (+ software pack). 123p.

A guide to the basics of astronomy for potential observers by Couper, with program software (suitable for the BBC model B and Electron computers) by Alpiar, which will allow the user to display the positions of Sun, Moon, planets and stars, as well as Halley's Comet, on a TV screen, for any time or place in the world. (From *Space Education* May 1985, p 432).

*164 Dickinson, Terence, Victor Costanzo and Glenn F. Chaple
THE EDMUND MAG 6 STAR ATLAS
Barrington, NJ: Edmund Scientific, 1982, spiral-bound. 67p.

Contains 12 principal charts facing pages describing salient deep sky objects, variables and double stars in the region. Provides helpful hints and techniques for observation. (From *S&T* March 1983, p 242).

*165 Dickinson, Terence
THE EDMUND SKY GUIDE (2nd ed.)
Barrington, NJ: Edmund Scientific, 1982, pbk

166 Dunlop, Storm and Wil Tirion
ATLAS OF THE NIGHT SKY
Feltham: Newnes, 1984, ISBN 0-600-35113-0. 80p; illus & diags (mostly col.). Bibl (2).

An atlas and a highly illustrated guide to objects and phenomena to be seen, with a chart for plotting the positions of the planets, 1984–2000. Contains 12 pages of Northern and Southern star maps (by Tirion), epoch 2000.0, down to magnitude 6, and 42 pages of constellation maps with descriptive notes. Includes a list of British and American societies.

*167 Fulton, Ken
THE LIGHT-HEARTED ASTRONOMER
Milwaukee, WI: AstroMedia, 1984, ISBN 0-913135-01-1 pbk. 115p.

A guide to backyard astronomy which regards the hobby of astronomy as a 'jungle' and stresses the dedication required to master the techniques and survive the setbacks. The author, with a sense of humour, and appropriate language, provides a guide through the jungle, notably that of advertising and products. (From *S&T* April 1985, p 324).

168 Gribbin, John
ASTRONOMY FOR THE AMATEUR
London: Macmillan, 1976, ISBN 0-333-18806-3; New York: McKay, 1977, ISBN 0-679-20384-2. 92p; illus (some col.), diags. Bibl (1), index (1).

A profusely illustrated, but rather brief guide for the complete beginner. Includes chapters on photography and radio astronomy.

169 Lancaster-Brown, Peter
STAR AND PLANET SPOTTING: A FIELD GUIDE TO THE NIGHT SKY
Poole: Blandford, 1981, ISBN 0-7137-0655-4; 0-7137-1265-1 pbk (reprint of 1st ed. 1974). 155 + 8p; illus (some col.), diags. Bibl (1), indexes (2).

Despite the title the author hopes this book will appeal to the armchair astronomer, as well as to the more practically minded novice. Includes star charts and finder maps for constellations, with notes on observing, but not on equipment.

170 Levitt, I.M. and Roy K. Marshall
STAR MAPS FOR BEGINNERS (rev. ed.)
New York: Simon & Schuster (Fireside Books), 1983. ISBN 0-671-47258-5 pbk. 62p; diags. Q

This guide, first published in 1942, contains monthly star charts in the shape of Maltese crosses to aid orientation in relation to the night sky. The text is quite substantial, covering naked-eye and basic telescopic observation, a description of the objects in the Solar

System, and the mythology of the constellations.

*171 Levy, David H.
THE JOY OF GAZING: A PERSONAL GUIDE FOR A NEW OBSERVER
Montreal: Royal Astronomical Society of Canada, 1982, pbk. 67p. Bibl.

An experienced amateur astronomer outlines subjects to explore for the beginner. Provides suggestions for viewing the Sun, planets, asteroids, meteors, occultations, variable stars, and eclipses, as well as hunting for comets and Messier objects. (From *S&T* June 1983, p 514).

172 Martin, Martha Evans, (revised by Don Rice and Craig Foltz)
THE FRIENDLY STARS
New York: Van Nostrand, 1982, ISBN 0-442-21198-8 pbk. 140p; diags. Index (2).

Regarded by some as a classic, this guide was first published in 1904. As the title suggests the author has imbued the book with a more personal flavour than similar works. Includes a number of locating maps for individual constellations, but no charts of the whole sky. One chapter covers constellations of the Southern sky.

173 Mayall, Robert Newton
THE SKY OBSERVER'S GUIDE: A HANDBOOK FOR AMATEUR ASTRONOMERS (rev. ed.)
New York: Golden Press, 1977, ISBN 0-307-24009-6 pbk. 160p; illus & diags (mostly col.). Index (2).

A small, colourful, pocket-sized guide for complete beginners, including younger observers.

174 Meyer, Ben
STARWATCH
New York: Putnam (Perigee Books), 1984, ISBN 0-399-51008-7; 0-399-51009-5 pbk. 144p; diags. Bibl (1), glos (1), index (1).

A guide to naked-eye and telescopic observation which uses large, ungainly diagrams printed in dark blue. The main part of the book is a series of constellation maps, with labels in huge type for outdoor visibility at night. The author introduces the concept of 'starframes', viewing aids made from coathangers!

175 Mitton, Jacqueline and Simon
STAR ATLAS
London: Jonathan Cape, 1979, ISBN 0-224-01715-2 pbk; New York:

Crown, 1979, ISBN 0-517-53737-0 pbk. 32p; illus, diags. Q

This basic guide is intended primarily for younger readers. Introductory chapters provide background information on astronomy and observation. The 14-page star atlas shows stars down to magnitude 6·5. Includes lists of variable stars, constellations and nebulae.

176 Moore, Patrick
STARGAZING: ASTRONOMY WITHOUT A TELESCOPE
London: Aurum Press, 1985, ISBN 0-906053-92-7; Woodbury, NY: Barron's Educational Series, 1985, ISBN 0-8120-5644-2. 176p + 4 charts; diags. Index (2). Q

A thorough guide for the complete beginner. Introductory chapters cover the nature of astronomy, the changing sky, constellations, star legends, and the nature of stars. The main part of the book is a series of basic monthly star charts for the Northern and Southern skies; for each month one looking north, another looking south. Coverage of the Northern sky is more thorough, with additional notes and constellation finder maps. The book concludes with notes on naked-eye objects in the Solar System. There are four separate glow-in-the-dark star charts in a pocket inside the back cover.

*177 Moore, Patrick
STARS OF THE AUSTRALIAN AND NEW ZEALAND SKIES
Windsor, Victoria: Currey O'Neil, 1980, ISBN 0-85550-344-0. 159p.

A revised version of an earlier work, *The Southern skies*.

178 THE NIGHT SKY
London: Times Books, annual. 28p.

Contains monthly star maps, with notes on the positions of the planets and the Moon. The maps are drawn for the latitude of London, and are published in *The Times* during the relevant year.

179 Raymo, Chet
365 STARRY NIGHTS
Englewood Cliffs, NJ: Prentice-Hall, 1982, ISBN 0-13-920520-9; 0-13-920512-8 pbk. 225p; illus, diags. Bibl (1), glos (2). Q

A year-long daily viewing programme for the beginning naked-eye observer. The text is fairly brief, and the general principles outlined in the book are rather scattered.

180 Ronan, Colin
THE PRACTICAL ASTRONOMER
London: Macmillan, 1981, ISBN 0-333-31245-7; London: Pan, 1981,

ISBN 0-330-26231-9 pbk; New York: Macmillan, 1981, ISBN 0-02-604500-1. 208p; illus & diags (some col.). Bibl (1), glos (1), index (2). Q

A guide to observation and practical work in astronomy, probably best suited to younger readers. Covers, in short sections, all types of objects and phenomena, with exercises, ending with astrophysical topics, such as the expanding universe and quasars, which are only described. The author's *Deep space* (37) is a companion volume.

181 Rükl, Antonín, (Olga Kuthanová, trans., John Gribbin, consultant ed., revised by Heather Couper)
AMATEUR ASTRONOMER (rev. ed.)
London: Peerage Books, 1985, ISBN 1-85052-001-1 (trans. of 1st Czech ed. 1979, revised). 192p; illus & diags (mostly col.). Bibl (1), index (1). Q

A guide to observational astronomy from naked-eye to basic telescopic work. The brief introductory discussion is followed by a long pictorial section (over 150 pages), which contains basic descriptive information facing colour artwork. Ends with notes on telescopes and photography.

*182 Schaaf, Fred
WONDERS OF THE SKY: OBSERVING RAINBOWS, COMETS, ECLIPSES, THE STARS AND OTHER PHENOMENA
New York: Dover, 1983, ISBN 0-486-24402-4 pbk; Magnolia, MA: Peter Smith, 1984, ISBN 0-8446-6099-X. 299p; illus.

An elementary guide for the general reader, intended partly for the beginning naked-eye observer and partly as a short introductory text. (From *S&T* Nov. 1984, p 421).

183 SIGNPOSTS TO THE STARS (4th ed.)
London: George Philip, 1982, ISBN 0-540-00214-3 pbk (reprint of 4th ed. 1970). 32p.

A basic guide to the identification of the constellations and the recognition of stars, with notes on the Sun and planets.

184 SKY CALENDAR
(Abrams Planetarium, Michigan State University, East Lansing, MI 48824-1324, USA), monthly, ISSN 0733-6314. 1 sheet; diags.

Intended for naked-eye observers of all ages. One side takes the form of a calendar, with diagrams and notes in the day boxes to help the user follow the motions of the Moon and planets. The other side features a simplified sky map, with brief notes.

185 STARS AT A GLANCE: A SIMPLE GUIDE TO THE HEAVENS (2nd ed.)
London: George Philip, 1982, ISBN 0-540-00335-2 pbk (reprint of 5th rev. impression, 1970, of 2nd ed. 1959). 52p.

A pocket-sized collection of monthly and seasonal star charts, with notes on constellations and stars.

186 Whitney, Charles A.
WHITNEY'S STAR FINDER (3rd ed.)
New York: Knopf, 1981, ISBN 0-394-74953-7 pbk; London: Michael Joseph, 1982, ISBN 0-7181-2129-5 pbk. 102p; diags.

One of the better guides to the night sky (not only to the stars). The 'star finder' (or, 'planisphere', as it is more often called) is double sided, showing both Northern and Southern stars. By rotating the appropriate side it is possible to see a representation of the sky at any particular time of the year. The book has chapters on the Moon, the planets, time, sunset and sunrise, eclipses, and the unpredictable, which highlight what can be seen, and give hints and ideas for practical work. Suitable for complete beginners, young and old.

187 Worvill, Roy
STARS AND TELESCOPES FOR THE BEGINNER
London: Kahn & Averill, 1979, ISBN 0-900707-50-X; New York: Taplinger, 1980, ISBN 0-8008-4464-5 pbk. 102p; diags. Glos (2), bibl (2).

This brief guide provides background astronomical information, advice on choosing and maintaining a telescope, and guidance for observing the Sun, Moon, planets and stars.

More detailed

Manuals for the serious amateur are listed in the section: Amateur Observing I, *General/manuals* (219–232).

188 Baker, David
THE HAMLYN GUIDE TO ASTRONOMY (UK 1978)/THE LAROUSSE GUIDE TO ASTRONOMY (US)/THE *COUNTRY LIFE* GUIDE TO ASTRONOMY (UK 1982)
London: Hamlyn, 1978, ISBN 0-600-33987-4; 0-600-30348-9 pbk (reprinted 1982); New York: Larousse, 1978, ISBN 0-88332-095-9 pbk. 288p; illus & diags (mostly col.). Glos (1), bibl (1), index (2).

A pocket guide, attractively illustrated to catch the casual reader, which provides an introduction to astronomy, including background physics, for the potential amateur observer. Includes a large section devoted to observation in which general star charts are followed by finder maps of individual constellations in clear diagrammatic form, with descriptive notes. Artwork by David Hardy.

189 Chartrand, Mark R., III and Helmut K. Wimmer
SKYGUIDE: A FIELD GUIDE FOR AMATEUR ASTRONOMERS (US)/AMATEUR ASTRONOMY: POCKET SKYGUIDE (UK)
New York: Golden Press, 1982, ISBN 0-307-13667-1 pbk; Feltham: Newnes, 1984, ISBN 0-600-35708-2 pbk. 272p; illus (mostly col.), diags. Bibl (2), index (2).

A colourful, highly illustrated guide for the beginning observer. Introductory chapters cover spherical astronomy, instruments, the atmosphere, stars, deep sky objects, and the Solar System. Includes star charts and constellation maps with features of interest described.

*190 Dickinson, Terence
NIGHTWATCH: AN *EQUINOX* GUIDE TO VIEWING THE UNIVERSE
Camden East, Ontario: Camden House, 1984, ISBN 0-920656-26-9 spiral-bound. 160p; illus, diags. Bibl. Q

An observing guide for the beginner and the slightly more advanced amateur. Begins with an introduction to the night sky, then presents seasonal charts in two versions: without colour or labels, and on the facing page in schematic form with objects named. Includes 20 atlas charts which give data in red, and, in blue, apparent magnitudes and notes on what might be seen in binoculars or a telescope. Also includes advice on selecting instruments, notes on celestial objects and phenomena, and a guide to resources for the amateur. (From *S&T* Jan. 1985, p 32–3).

191 Dunlop, Storm
ASTRONOMY: A STEP BY STEP GUIDE TO THE NIGHT SKY
Feltham: Hamlyn, 1985, ISBN 0-600-30557-0 pbk; New York: Collier Macmillan, 1985, ISBN 0-02-079650-1 pbk. 192p; illus & diags (mostly col.). Bibl (1), index (2).

A concise, but not superficial guide. The first part provides essential information for the beginner, including notes on drawing (unusual in such books) and photography, naked-eye observation, binoculars and telescopes, star charts, and a guide to the constellations. Part two describes features to look for in the Solar System and beyond.

*192 Kals, William S.
THE STARGAZER'S BIBLE
New York: Doubleday, 1980, ISBN 0-385-13057-0 pbk. 168p; illus.

This introduction includes sections on identifying constellations, finding bright stars and planets, stargazing at other latitudes, binoculars, telescopes, astrophotography, and concludes with a useful appendix of observational data. (From *Mercury* May/June 1983, p 89).

193 Klepešta, Josef and Antonín Rükl, (Olga Kuthanová, trans.)
CONSTELLATIONS: A CONCISE GUIDE IN COLOUR
London: Hamlyn, 1978, ISBN 0-600-00893-2 (reprint of trans. of 1st Czech ed. 1969). 282p; col. illus, diags. Glos (3).

The main part of this book is a set of charts of individual constellations, alphabetically arranged, with explanatory text facing each one. Introductory chapters give background information and notes on observing, but not on equipment.

*194 Liller, Bill and Ben Meyer
THE CAMBRIDGE ASTRONOMY GUIDE: AN INTRODUCTION TO PRACTICAL ASTRONOMY
Cambridge: Cambridge University Press, 1985, ISBN 0-521-25778-6.

Intended for lovers of astronomy who wish to do more than just look at the night sky or marvel at glossy pictures of it. Tells how to take star photographs with nothing more than a simple camera. Meyer's odd-numbered chapters explain how one novice quickly progressed to become one of the best-known amateur astronomers. Liller's even-numbered chapters comment on Meyer's account from the professional point of view. (Publisher's information).

195 Menzel, Donald H. and Jay M. Pasachoff
A FIELD GUIDE TO THE STARS AND PLANETS (3rd ed.)
Boston: Houghton Mifflin, 1983, ISBN 0-395-34641-X; 0-395-34835-8 pbk. 475p; illus (some col.), diags. Glos, bibl (2), index.

One of the better guides, this one is intended specifically for observers wishing to learn their way about the night sky. It is designed for use in both the Northern and Southern hemispheres. Includes notes on observing the Moon (with maps), the Sun, the planets and other objects in the Solar System. Covers telescopes and their use, photography, and time. There are numerous tables of data and objects for observation. (Based on content of 2nd ed. 1975). The third edition has been completely rewritten by Pasachoff, with monthly sky maps and 52 detailed atlas charts by Wil Tirion, covering the entire sky down to stars of magnitude 7·5. (From *S&T* Sept. 1984, p 227–8).

196 Moore, Patrick
THE NEW OBSERVER'S BOOK OF ASTRONOMY (6th ed.)
London: Frederick Warne, 1978, ISBN 0-7232-1646-0 pbk (reprinted in a new format 1983). 192 + 32p; illus (some col.). Bibl (1), index (1).

A guide for beginners of all ages. Introduces the basic principles of astronomy and describes the techniques of star recognition. One chapter deals with basic equipment. (Previously titled *The Observer's book of astronomy*).

197 Moore, Patrick
PATRICK MOORE'S GUIDE TO THE STARS AND PLANETS (rev. ed.)
London: Mitchell Beazley, 1982, ISBN 0-85533-196-8. 144 + 16p; illus (some col.). Glos (1), index (1).

'A working manual for serious amateur observers all over the world' (at the start of their careers). Divided into two main sections: the Solar System, and the stars, which provide descriptive information. The introduction covers basic understanding, telescopes and observing. Includes lunar maps and star charts, with indexes.

*198 Muirden, James
ASTRONOMY WITH A SMALL TELESCOPE
London: George Philip, 1985, ISBN 0-540-01088-X. 256p; illus, diags. Bibl, index.

Advice on what to see and where to look. The core of the book is a month by month guide, illustrated with detailed star charts for observers in both the Northern and Southern hemispheres. (Publisher's information).

199 Muirden, James
ASTRONOMY WITH BINOCULARS (2nd ed.)
London: Faber, 1976, ISBN 0-571-04903-6; 0-571-04904-4 pbk; New York: Crowell, 1979, ISBN 0-690-01723-5; New York: Arco, 1979, ISBN 0-668-05832-3 pbk. 173 + 8p (UK), 192p (US); illus, diags. Bibl (1), index (2).

This guide covers selection of instruments and what to observe, with a final chapter of observing programmes. The US edition includes an appendix on Halley's Comet.

*200 Ottewell, Guy
ASTRONOMICAL CALENDAR
Greenville, SC: Furman University Physics Dept., 1974–, annual, pbk. About 65p; diags. Q

This large-sized volume includes monthly star maps and sun-centred planet charts, with day-by-day notes on interesting sky events. The explanatory text provides useful observing information on (and includes finder charts for) planets, asteroids and notable comets. (From *S&T* Jan. 1983, p 42 and March 1985, p 229).

201 Ottewell, Guy
ASTRONOMICAL COMPANION
Greenville, SC: Furman University Physics Dept., 1979, ISBN 0-934546-01-0 pbk (revised and corrected, 1981). 73p; diags. Bibl (2). Q

A compilation of the non-year-related material from the author's *Astronomical Calendar* (200), for explanation, reference or enjoyment. Contains many well drawn diagrams, some using a three-dimensional approach for added clarity. Can be used as a supplement to a short non-mathematical astronomy course.

202 Rey, H.A.
THE STARS: A NEW WAY TO SEE THEM (rev. ed.)
Boston: Houghton Mifflin, 1967, ISBN 0-395-08121-1; 1976, 0-395-24830-2 pbk; London: Chatto & Windus, 1975, ISBN 0-7011-5051-3. 160p; illus, diags. Bibl (1), glos/index (2). Q

'This book [first published in 1952] is meant for people who want to know just enough about the stars to be able to go out at night and find the consellations, for the pleasure of it'. Despite this modest aim the book contains a not insignificant text, providing notes on position and movement, variations with time, the various celestial objects, with a catalogue of the constellations and star charts. Not listed by Seal (24).

203 Ridpath, Ian and Wil Tirion
COLLINS GUIDE TO STARS AND PLANETS (UK)/
UNIVERSE GUIDE TO STARS AND PLANETS (US)
London: Collins, 1984, ISBN 0-00-219071-0; 0-00-219067-2 pbk; New York: Universe Books, 1985, ISBN 0-87663-366-1. 384p; illus & diags (mostly col.). Index (2).

An attractively illustrated guide. The first section consists of monthly star charts and constellation maps, drawn by Tirion, with notes and some photographs. The second section (about 120 pages) is a highly illustrated basic introduction to astronomy.

204 Ronan, Colin (consultant ed.), (revised and updated by Storm Dunlop)
AMATEUR ASTRONOMY: A COMPREHENSIVE AND PRACTICAL SURVEY (UK)/THE GOLDEN BOOK OF ASTRONOMY (US)
Feltham: Newnes, 1984, ISBN 0-600-35667-1; Racine, WI: Western Publishing Co., 1984, ISBN 0-307-46649-3. 256p; illus & diags (some col.). Glos (2), bibl (2), index (2). Q

A highly illustrated introduction to astronomy with a very strong emphasis on the practical side. The early chapters deal with motions of celestial bodies, the stars, the Sun, the Solar System, the Galaxy, and extragalactic astronomy, each having one or more sections of exercises. The final chapters cover cosmology and professional astronomical observation. Includes ten pages of star charts.

205 Ronan, Colin A. (consultant ed.), Storm Dunlop and Brian Jones
THE SKYWATCHER'S HANDBOOK
New York: Crown, 1985, ISBN 0-517-55703-7; London: Transworld (Corgi), 1985, ISBN 0-552-99196-1; 0-552-99146-5 pbk. 224p; illus & diags (mostly col.). Glos (1), index (2).

An attractively illustrated guide to all skywatching, not only the astronomical variety. The book's three parts cover: 1. the daylight sky, 2. the night sky, and 3. observing and recording the sky. Includes basic star maps for Northern and Southern observers. The appendices provide data such as planetary positions 1985–2000, and dates of key celestial events.

Amateur Observing I

Reference

For specific reference works *see* (233–247, stellar and deep sky) and (265–275, Solar System).

206 ALMANAC FOR COMPUTERS
Washington, DC: Nautical Almanac Office, annual, pbk. Q

Provides data in a form to facilitate the application of small computers and calculators to navigational and positional astronomy.

207 THE ASTRONOMICAL ALMANAC
London: HMSO, 1767–; Washington, DC: US Government Printing Office, 1855–, annual. Glos (3), index (2).

Produced jointly by the Nautical Almanac Office at the Royal Greenwich Observatory and the US Naval Observatory, this is a compendium of general data vital for the professional, but also very useful for the student and the serious amateur. Includes tables of co-ordinates and observing data for the main members of the Solar System, lists of stars, galaxies, radio and x-ray sources, pulsars, and orbital elements of minor planets. There is much information on time scales and co-ordinate systems. Includes an index of observatories. (The British *Astronomical ephemeris* (which had replaced *The nautical almanac and astronomical ephemeris* in 1960) and the *American ephemeris and nautical almanac* were amalgamated in 1981).

208 ASTRONOMICAL PHENOMENA FOR THE YEAR . . .
Washington, DC: US Government Printing Office; London: HMSO, annual, pbk.

Published two years in advance, this booklet is a useful digest of information in (207). Includes such data as the phases of the Moon, eclipses, occultations, perigee and apogee of the Moon, the visibility of the planets, a diary of configurations of the Sun, Moon and planets, tables of sunrise and sunset, and moonrise and moonset.

209 AN ASTRONOMICAL CALENDAR
Cardiff: Cardiff Astronomical Society and the Federation of Astronomical Societies, annual, pbk. About 45p; diags.

A guide to the year's celestial events which provides background information and explanations, so is suitable for the inexperienced observer. Includes monthly star charts, charts of planetary positions, information on the larger minor planets, meteor showers, eclipses and notable comets. (Write to: D.W.S. Powell, 1 Tal y Bont Road, Ely, Cardiff, CF5 5EU, UK).

210 ASTRONOMICAL HANDBOOK FOR SOUTHERN AFRICA
Cape Town: The Astronomical Society of Southern Africa, annual, ISSN 0571-7191 pbk. About 50p; diags.

'Intended both as an introduction to observational astronomy for the interested layman . . . and as a handbook for the established amateur or professional astronomer'. Includes a short introduction on astronomy in Southern Africa, useful data and tables of phenomena, with explanations and descriptive notes.

211 THE HANDBOOK OF THE BRITISH ASTRONOMICAL ASSOCIATION
London: British Astronomical Association, 1921–, annual, pbk. About 115p; diags. Ads.

A compendium of useful information for the serious amateur observer, some of which is taken from *The Astronomical Almanac* (207). Universal Time is used throughout. Includes rising and setting times of planets, radio time signals, moonrise and moonset, lunar occultations, planetary positions throughout the year, predicted cometary positions, a meteor diary, stellar data, and physical data of the Sun, Moon, planets and satellites.

*212 Mansfield, Roger
SKYWATCHER'S ALMANAC
Colorado Springs, CO: Astronomical Data Service, annual.

Information about stars, planets and eclipses (both solar and lunar), with a 12-page computer printout for the purchaser's own geographical location. (Write to: 3922 Leisure Lane, Colorado Springs, CO 80917, USA). (From *S&T* March 1984, p 238).

*213 Mansfield, Roger
SUNLIGHT SUMMARY
Colorado Springs, CO: Astronomical Data Service, 1979, pbk. 105p.

A computer-generated table of local sunrise, sunset, and twilight for each day of the year, for the purchaser's latitude and longitude, together with extensive background information. (Address: *see* 212). (From *Mercury* May/June 1981, p 92).

214 Moore, Patrick
THE GUINNESS BOOK OF ASTRONOMY FACTS AND FEATS (2nd ed.)
Enfield: Guinness Superlatives, 1983, ISBN 0-85112-258-2; 0-85112-294-4 pbk. 289 + 8p; illus (some col.), diags. Glos (2).

Despite the typical 'Guinness Book' presentation of 'biggests' and 'bests' this book contains much useful data for the beginning, or less experienced observer. Includes separate sections for each body, or groups of bodies in the Solar System, with numerous tables and facts of all kinds. Stars and deep sky objects, and the constellations are covered in a similar fashion. The star catalogue of nearly 100 pages has been revised for epoch 2000.0 in this edition, and stellar distances are now given in parsecs. Features chronological listings showing the development of astronomy and space science, and biographical sketches of astronomers no longer living.

215 Moore, Patrick
YEARBOOK OF ASTRONOMY
London: Sidgwick & Jackson; New York: Norton, 1962–, annual, (hardcover & pbk). Illus, diags.

Each yearly volume follows a similar format: monthly star charts, with descriptive notes, in the first section, followed by articles on various subjects, of varying degrees of technicality, mostly reporting on the latest research, but with a fair amount of historical material and information for the amateur. The third part consists of tables of interesting deep sky objects for the observer. Ends with a list of astronomical societies.

216 OBSERVER'S HANDBOOK
Toronto: Royal Astronomical Society of Canada, 1907–, annual, ISSN 0080-4193 pbk. About 180p; diags. Bibl (1), index (1).

A compendium of useful information for the amateur and the professional astronomer. Similar to the BAA handbook (211), but with more descriptive material. Includes a detailed listing of the year's celestial events. Unlike the BAA publication this handbook has lunar maps and (bi-monthly) star charts. (Before 1911 the title *Canadian astronomical handbook* was used).

217 Robinson, J. Hedley and James Muirden
ASTRONOMY DATA BOOK (2nd ed.)
Newton Abbot: David & Charles, 1979, ISBN 0-7153-7743-4; New York: Halsted Press, 1979, ISBN 0-470-26594-9. 272p; diags. Glos (3), index (2).

'Intended as a reference tool for the student and amateur astronomer, and for those interested in the earth sciences'. Brings together scattered information from various publications. Should be used in conjunction with a good star atlas. In addition to the physical data, much of which is in tabular form, there is a chronology, and a section on telescopes. There are wide margins and blank areas for notes.

218 Satterthwaite, Gilbert E. (ed.)
NORTON'S STAR ATLAS AND REFERENCE HANDBOOK (EPOCH 1950.0) (17th ed.)
Edinburgh: Gall & Inglis, 1978, ISBN 0-85248-900-5. 118 + 35p;

diags. Index (2), glos (2). Q

The amateur astronomer's bible, first published in 1910, is also aimed at the general reader as a reference book. This edition has been completely reset, with metric units used throughout. The first five sections comprise the reference handbook, and cover background physics, instrumentation, and observation; useful tables; the Solar System, including maps of the Moon and Mars; stars, nebulae and galaxies; telescopes and accessories. The final section of star charts shows stars down to magnitude 6·35, and includes lists of interesting objects.

General/manuals

219 THE ASTRONOMER
(John A. Colls, 177 Thunder Lane, Thorpe St. Andrew, Norwich, NR7 0JF, UK), 1964–, monthly. Illus, diags. Q

'One of the world's leading preliminary observational report magazines', it is produced by, and intended for active amateur astronomers. Features drawings, charts and photographs on the covers. The rest of the magazine, which is duplicated from typescript, contains the reports, divided into sections covering comets, variable stars, asteroids, meteors, the planets, and the Sun. Aims to publish all observable astronomical objects. Includes articles relating to photography and observational techniques. Regularly features ephemerides for new and recovered comets, and predictions for important asteroids. The 'news notes' provide updates on the latest discoveries made by amateurs and professionals. The magazine also runs a number of services for amateurs, involving a number of bulletins and circulars, all available for a supplementary subscription: *Asteroidal Bulletins, Atlas Omission Lists, Early Warning Circulars, Nova/Supernova Circulars, Observers' Bulletins,* and *Variable Star Circulars*.

220 Baird, Michael
WEATHER FORECASTING FOR ASTRONOMY
Milford, CT: Winmark Press, 1982, ISBN 0-9608278-0-3 pbk. 107p; illus, diags.

The aim of this book is to increase the understanding of weather forecasting of amateur astronomers. No previous meteorological knowledge is assumed. The author argues that observers can take specific action to improve their seeing by taking advantage of the fact that there will be significant variations in the weather of a relatively small area. Discusses how particular geographical locations shape the weather at night.

*221 Fox, James H.
MATH FOR AMATEUR ASTRONOMERS
Sioux City, IA: Astronomical League, 1983, pbk. 59p; diags. Bibl.

The usefulness of this text lies in its presentation of a concise summary of the basic mathematics that the amateur can use in observing, as well as that which applies to the telescope itself. It is suitable for amateurs with a background of algebra and a little trigonometry. Does not cover photometry, radio astronomy and spectroscopy. (From *The Reflector* Nov. 1983, p 6).

*222 Green, Robin M.
SPHERICAL ASTRONOMY
Cambridge: Cambridge University Press, 1985, ISBN 0-521-23988-5; 0-521-31779-7 pbk.

An introduction to spherical and positional astronomy suitable for use by students with no prior knowledge of classical astronomy. (Publisher's information).

223 IAU CIRCULARS
Cambridge, MA: Central Bureau for Astronomical Telegrams, Smithsonian Astrophysical Observatory. About 120/year.

Postcard-sized notifications of information about astronomical phenomena that requires prompt dissemination. The service is used by most of the world's major observatories, and by many leading amateurs.

224 JOURNAL OF THE BRITISH ASTRONOMICAL ASSOCIATION
(British Astronomical Association, Burlington House, Piccadilly, London, W1V 0NL, UK), 1890–, bi-monthly. Illus, diags. Ads. Q

The chief forum in print for amateur astronomers in the UK. Includes articles at varying levels on most aspects of astronomy, but strongly slanted toward the practical. Carries reports on observations and Association matters, as well as book reviews and correspondence. The BAA also produces the *Memoirs*, issued irregularly, which contain longer reports of the various observing sections, and *Circulars*, which bring immediate information of newly discovered objects, such as novae and comets.

The sections of the BAA are listed below; some produce, usually irregular, periodicals of their own; these appear in brackets: Artificial Satellite, Aurora *(Newsletter)*, British Fireball Survey, Comet *(Bulletin)*, Computing (this section is responsible for the *Handbook* (211)), Deep Sky *(Deep Sky Diary)*, Historical, Jupiter, Lunar *(Circular, The New Moon*, also *Guide for observers of the*

Moon, 3rd ed. 1979), Meteor *(Newsletter)*, Observing techniques (a series of *Observational Notes*), Saturn, Solar *(Solar Bulletin, Solar Summary)*, Terrestrial Planets *(Inner Planets Newsletter, Notes)*, Variable Star *(Circular: 'Light Curve'*, ISSN 0267-9272). (*See also* 610).

225 Muirden, James
THE AMATEUR ASTRONOMER'S HANDBOOK (3rd ed.)
New York: Harper & Row, 1983, ISBN 0-06-181622-1. 427p; illus, diags. Glos (2), bibl (2), index (3).

This book for the serious amateur observer assumes a certain amount of basic astronomical knowledge. It is intended as 'a survey of the technique of amateur astronomy, from the selection of an instrument to the conduct of actual observation'. The five sections cover equipment, the Solar System, the stars and nebulae, astrophotography, and optical work for amateurs. Includes an annotated bibliography.

226 Robinson, Leif L. (ed.)
THE *SKY & TELESCOPE* GUIDE TO THE HEAVENS
Cambridge, MA: Sky, 1980, ISBN 0-933346-29-8 pbk. 52p; illus, diags. Bibl (1). Q

Designed as a companion to the *AAVSO variable star atlas* (244), this volume of reprints deals with more than just variable stars, covering also comets, asteroids, and deep sky objects. The first half of the book is devoted to variable stars, and includes an index by constellation to the AAVSO atlas.

227 Roth, G.D., (Arthur Beer, trans. & rev.)
ASTRONOMY: A HANDBOOK
New York: Springer-Verlag, 1975, ISBN 0-387-06503-2 (based on 2nd German ed. 1967), (reprinted Cambridge, MA: Sky, no other details known). 567p; illus, diags. Bibl (3), index (3).

This rather demanding volume is essentially for amateurs, but is also designed to meet the needs of professionals. The emphasis is on instruction over the presentation of facts: the content is both theoretical and practical. Elementary astronomical, mathematical and physical knowledge is assumed. The topics covered are observing instruments, radio astronomy for amateurs, fundamentals of spherical astronomy, sundials, applied mathematics for amateurs, observation of objects in the Solar System including artificial satellites, stars (photometry, double stars), the Milky Way and other galaxies.

228 Sherrod, P. Clay with Thomas L. Koed
A COMPLETE MANUAL OF AMATEUR ASTRONOMY: TOOLS AND TECHNIQUES FOR ASTRONOMICAL OBSERVATIONS
Englewood Cliffs, NJ: Prentice-Hall, 1981, ISBN 0-13-162115-7; 0-13-162107-6 pbk. 319p; illus, diags. Bibl (x), glos (3), index (2).

This guide for the serious beginner is not intended as a broad overview of astronomy so many subjects have been purposely omitted. The emphasis is on research projects (unlike similar guides), showing how the amateur can made a meaningful contribution to the science of astronomy. Covers telescopes, instrument setup and maintenance, observation of meteors, comets and the Sun, lunar topography and lunar occultations, Mars, Jupiter, Saturn, visual photometry of the minor planets, variable stars, an introduction to photoelectric photometry, and astrophotography. Among the 15 appendices are examples of observing forms for various objects and events, and guidance on organizing an astronomy club.

229 Sidgwick, J.B., (revised by James Muirden)
AMATEUR ASTRONOMER'S HANDBOOK (4th ed.)
London: Pelham Books, 1979, ISBN 0-7207-1164-9; Hillside, NJ: Enslow, 1980, ISBN 0-89490-049-8; 0-89490-076-5 pbk. 568p; diags. Bibl (3), index (3).

The standard work for the serious amateur. A compendium of information covering the theoretical and instrumental background to observation. Covers the function of an astronomical telescope, optics, reflecting and refracting telescopes, testing, mounting and using telescopes, accessories, photography, photometry, vision (including physiology of the eye), the atmosphere and 'seeing', time and clocks, and the mathematical treatment of observations. The ten appendices cover such matters as keeping records, observational aids, (British) manufacturers and suppliers.

This book should not be confused with the reprint of the third edition, revised by R.C. Gamble, 1971 (New York: Dover, 1980, ISBN 0-486-24034-7 pbk).

230 Sidgwick, J.B., (revised by James Muirden)
OBSERVATIONAL ASTRONOMY FOR AMATEURS (4th ed.)
London: Pelham Books, 1982, ISBN 0-7207-1378-1; Hillside, NJ: Enslow, 1982, ISBN 0-89490-067-6; 0-89490-068-4 pbk. 348p; diags. Bibl (3), index (2).

A companion to (229), this volume deals with observational techniques, covering the major bodies of the Solar System individually, and including selected data from the BAA handbook (211) and *The astronomical almanac* (207). Further chapters cover asteroids, zodiacal light, aurorae, meteors, comets, variable stars, binary stars, nebulae and clusters.

This book should not be confused with the reprint of the third edition, revised by R.C. Gamble, 1971 (New York: Dover, 1980, ISBN 0-486-24033-9 pbk).

*231 van Zandt, Rollin P.
ASTRONOMY FOR THE AMATEUR, VOL. 1: PLANETARY ASTRONOMY (3rd ed.)
Peoria, IL: R.P. van Zandt, 1977, looseleaf. 243p.

An introductory text for classes in astronomy conducted by amateur societies. The text is reproduced from typescript; the pages are punched for looseleaf binding, the chapters being individually paginated to facilitate the insertion of updating sheets. (From *S&T* Dec. 1977, p 518).

*232 van Zandt, Rollin P.
ASTRONOMY FOR THE AMATEUR, VOL. 2: STELLAR ASTRONOMY (3rd ed.)
Peoria, IL: R.P. van Zandt, 1977, looseleaf. 257p.

This companion to (231), in the same format, provides information on stars, clusters, nebulae, the Milky Way and other galaxies. Concludes with an extended section on astronomical instruments. (From *S&T* Jan. 1979, p 84).

Amateur Observing II — Stellar and Deep Sky

Reference

233 BROWN'S STAR ATLAS (new ed.)
Glasgow: Brown, Son & Ferguson, 1977, ISBN 0-85174-271-8. 49p; diags. Q

Essentially for use in navigation at sea. Part one, for beginners, has six large charts, each showing a section of the sky which overlaps the next section. Smaller maps help to identify individual constellations. Part two covers the problems of stellar navigation. No index, but the contents page is fairly detailed.

*234 Coutchie, Raymond P.
DEEP SKY CATALOG AND STAR ATLAS
Woodland Hills, CA: R.P. Coutchie, 1983, pbk (reprint of 1st ed. 1963). 17p.

Eight star charts covering the sky to 5th magnitude face tables listing interesting objects. (From *S&T* Aug. 1984, p 136).

*235 Coutchie, Raymond P.
MESSIER: A COMPLETE CATALOG
Woodland Hills, CA: Diversified Design, 1983, pbk. 28p.

A guide organized by right ascension which provides a small- and large-scale chart for each object along with a brief description and blank space for notes. (From *S&T* June 1984, p 527).

236 Freeman, Lenore
A STARHOPPER'S GUIDE TO THE MESSIER OBJECTS
Oakland, CA: Everything in the Universe, 1983, ISBN 0-913399-57-4 pbk. 28p; diags. Q

A set of simple constellation maps showing the Messier objects, with a table listing data such as right ascension and declination, type of object, visibility, and distance.

237 Hirshfeld, Alan and Roger W. Sinnott
SKY CATALOGUE 2000.0 VOL. 1: STARS TO MAGNITUDE 8·0
Cambridge: Cambridge University Press, 1982, ISBN 0-521-24710-1; 0-521-28913-0 pbk; Cambridge, MA: Sky, 1982, ISBN 0-933346-35-2; 0-933346-34-4 pbk. 604p. Q

Intended for use in conjunction with Wil Tirion's *Sky atlas 2000.0* (245), this catalogue presents tabular data on 45,269 stars of visual magnitude 8·05 and brighter. It is based on data 'derived from the character-coded computer tape of SKYMAP, version 3.0', prepared at the NASA Goddard Space Flight Center. The stars are listed in RA order, giving declination, proper motion, visual magnitude and colour index in the UBV photometric system, absolute visual magnitude, spectral type, radiant velocity, and distance in parsecs. Includes indexes to named stars. A major reference for the serious amateur.

238 Hirshfeld, Alan and Roger W. Sinnott
SKY CATALOGUE 2000.0 VOL. 2: DOUBLE STARS, VARIABLE STARS AND NONSTELLAR OBJECTS
Cambridge: Cambridge University Press, 1985, ISBN 0-521-25818-9; 0-521-27721-3 pbk; Cambridge, MA: Sky, 1985, ISBN 0-933346-39-5;

0-933346-38-7 pbk. 57 + 385p; illus, diags. Bibl (3), Glos (2).

This volume contains 15 main data sections covering double and multiple stars, variables, open clusters (including a cross index to other catalogues), globular clusters, nebulae, galaxies, quasi-stellar objects, radio and x-ray sources. The introduction provides notes on the different objects, and includes an index to letter names of variable stars. The authors express the hope that they have provided more than just a list of numerical data, but a collection of observing challenges for amateur astronomers.

*239 Klein, Fred W.
A NEW LIST OF 450 DEEP-SKY OBJECTS FOR AMATEUR TELESCOPES
Hawaiian Volcano Observatory, HI: F.W. Klein, 1981, pbk. 31p.

A reference list of objects down to about 9th magnitude. Includes nebulae, star clusters, galaxies, double stars and more, with celestial co-ordinates. (From *S&T* Dec. 1982, p 559).

240 Klein, Fred
THE VISIBILITY OF DEEP-SKY OBJECTS; WITH DATA ON OVER 500 OBJECTS (5 volume set; VOL. 1: 564 BASIC TARGETS FOR SMALL TELESCOPES)
Los Altos, CA: Klein Publications, 1983, ISBN 0-913051-02-0 (5 vol. set). 283p. ISBN 0-913051-03-9 (vol. 1) pbk. 59p.

Volume one of the five-volume set introduces the concept of 'visibility classes' for astronomical objects, allowing the amateur to determine how easily a particular object can actually be seen with a small telescope. Lists the basic objects by RA, by type and by visibility, ending with a list of the author's 50 most impressive objects. Volumes two and three cover 2,400 objects (including 1,270 galaxies), volumes four and five cross reference the lists with Wil Tirion's *Sky atlas 2000.0* (245).

*241 Maxwell, Russell J. and Rollin P. van Zandt
THE STELLAR HANDBOOK
Carpentersville, IL: Astronomical League, 1977, pbk. 56p.

Arranged by right ascension for 1980 positions, this reference listing by deep sky objects for amateur observers gives extensive basic data about single and multiple stars, nebulae, clusters, and galaxies. (From *S&T* Feb. 1978, p 170).

*242 Salmi, Juhani
CHECK A POSSIBLE SUPERNOVA
Lahti, Finland: J. Salmi, 1984, spiral-bound. 40 cards.

A compilation of 40 galaxy photographs for use at the telescope on handy [13·5 × 10cm] cards. Visual observers can check stars against photographs to search for a supernova. The images are negative (black stars on a white background). (From *S&T* Sept. 1984, p 230). (Available from AstroMedia Corp. in the US, and the Webb Society in the UK, *see* 135 and 264 for addresses).

*243 Sarna, Thomas
SARNA DEEP SKY ATLAS
Richmond, VA: Willmann-Bell, 1984. 102 charts + finder map. Q
A set of 102 very large scale charts (1 arc minute per mm.), showing white star fields on a dark blue background, with reference data at the top of each chart. Unlike most guides to the heavens which locate objects only by right ascension and declination, these charts also provide simple directions for aiming a telescope at the right spot by 'offsetting' from bright, easily located stars. (From *S&T* Feb. 1985, p 135).

*244 Scovil, Charles E.
AAVSO VARIABLE STAR ATLAS
Cambridge, MA: Sky, 1980. Boxed set of 178 sheets.
An atlas for serious variable star observers, adapted from the now out-of-print *Smithsonian Astrophysical Observatory star atlas*, which shows the sky from pole to pole to about magnitude 9·5, using 1950.0 co-ordinates. Includes over 2,000 variable stars, many with comparison star magnitudes surrounding them, 1,249 galaxies brighter than 13th magnitude, and numerous other deep sky objects. Includes *The Sky & Telescope guide to the heavens* (226) as a supplement. (From *S&T* March 1981, p 245).

245 Tirion, Wil
SKY ATLAS 2000.0 (deluxe ed.)
Cambridge: Cambridge University Press, 1981, ISBN 0-521-24467-6 ring-bound; Cambridge, MA: Sky, 1981, ISBN 0-933346-33-6 ring-bound. 26 foldout sheets + 8p; col. charts. Bibl (1). Q (39.5 × 27cm folded).
'The first comprehensive set of star charts for this new co-ordinate epoch' includes about 43,000 stars down to visual magnitude 8·0, and about 2,500 deep sky objects, 'compiled from some of the best current sources'. The map projections have been selected to avoid distortion, so that star fields are plotted as they appear in the sky. A useful feature for the less advanced observer is the labelling of objects with their popular names. Also available from the US publisher (Sky): desk ed. (black stars on a white background) ISBN

0-933346-31-X; field ed. (white stars on black) ISBN 0-933346-32-8. An article by Wil Tirion describing his work on this atlas appeared in *S&T* June 1981, p 504–6.

246 Vehrenberg, Hans
ATLAS OF DEEP-SKY SPLENDOURS (4th ed.)
Cambridge, MA: Sky, 1983, ISBN 0-933346-03-4; Cambridge: Cambridge University Press, 1984, ISBN 0-521-25834-0. 242p; illus (some col.), diags. Q

First published in 1967, this highly regarded work presents the brighter deep sky objects in 113 photographs in a uniform scale of one millimetre to one second of arc, with finder charts, co-ordinates and other data. Also includes close-up photographs of selected objects, with descriptions and commentaries.

247 Vehrenberg, Hans and Dieter Blank
HANDBOOK OF THE CONSTELLATIONS (4th ed.)
Düsseldorf, West Germany: Treugesell-Verlag, 1981. 197p; diags.

A spectacular combination of astronomical charts (white stars on a black background) and a catalogue (giving positions for epoch 1950.0). The catalogue pages, arranged by type of object, face the charts which are presented alphabetically by constellation and thumb indexed. Ends with an explanatory section. The text is in English and German, with astronomical names also in Latin.

General

248 Burnham, Robert, jr.
BURNHAM'S CELESTIAL HANDBOOK: AN OBSERVER'S GUIDE TO THE UNIVERSE BEYOND THE SOLAR SYSTEM (3 vols.)
New York: Dover, 1978, ISBN 0-486-23567-X (vol. 1) pbk; ISBN 0-486-23568-8 (vol. 2) pbk; ISBN 0-486-23673 (vol. 3) pbk; 1980, ISBN 0-486-24063-0 (vol. 1); ISBN 0-486-24064-9 (vol. 2); ISBN 0-486-24065-7 (vol. 3). 2138p; illus, diags. Bibl (2), indexes (2).

'Intended to be a standard catalog and detailed descriptive handbook to the many thousands of objects available to observers with telescopes in the 2-inch to 12-inch range'. The almost 100 pages of introductory matter provide background astronomical information for the beginner, and give instructions on the use of the book. The bulk of the three volumes is an alphabetical catalogue of constellations with finder charts, and, for each, lists of double and multiple stars, variable stars, clusters, nebulae and galaxies. The extensive decriptive notes often venture out of the field of astronomy, to art and mythology, for example. (Shorter versions of volumes two and three were published in 1966).

249 Couteau, Paul, (Alan Batten, trans.)
OBSERVING VISUAL DOUBLE STARS
Cambridge, MA: MIT Press, 1981, ISBN 0-262-03077-2 (trans. of 1st French ed. 1978). 257p; illus, diags. Bibl (x), index (2).

This enthusiastic and inspirational volume is aimed at anyone interested in the subject, from the young beginner to the experienced astronomer. The eight chapters, which are independent of each other, cover history, relevant optical concepts, measuring instruments, observational practice, identification of stars (with brief notes on the history of star catalogues). There is a long chapter intended for students on the computation of orbits and stellar masses, and a description of the process of astronominal research based on the author's work at Nice. Ends with a catalogue of 744 double stars (where the primary is almost always brighter than magnitude 8, and the secondary, magnitude 10), with notes. The references given are mostly technical and often French.

250 DEEP SKY
(AstroMedia Corp., 625 East St. Paul Avenue, Milwaukee, WI 53202, USA), 1982–, quarterly, ISSN 0735-3073. About 40p; illus, diags. Ads. Q

'The magazine for deep sky observers and astrophotographers' contains notes for observers of double stars, multiples, variables, nebulae, galaxies, and quasars; information on equipment and techniques; feature articles; correspondence, book reviews, and reports on the professional literature. (This is a successor to *Deep Sky Monthly*, 1977–82).

*251 Guzman, L., B. Guzman and P. Jones
OBSERVE THE HERSCHEL OBJECTS
Carpentersville, IL: Astronomical League, 1980 pbk. 40p.

An observing guide to 400 of the almost 2,500 celestial objects observed and catalogued by William Herschel, which are accessible to 6-inch, or larger telescopes. (From *Mercury* May/June 1982, p 98).

252 Hartung, E.J.
ASTRONOMICAL OBJECTS FOR SOUTHERN TELESCOPES: A HANDBOOK FOR AMATEUR ASTRONOMERS
Cambridge: Cambridge University Press, 1984, ISBN 0-521-31887-4

pbk (reprint of 1st ed. 1968). 238 + 16p; illus. Bibl (2), index (1).

This guide begins with general information and notes on amateur observing, followed by tables of 1,017 objects in right ascension order. The bulk of the book is the descriptions of each object alphabetically arranged by constellation. Also lists selected objects for Northern observers.

253 Jones, Kenneth Glyn (ed.), [Robert Argyle]
WEBB SOCIETY DEEP-SKY OBSERVER'S HANDBOOK, VOL. 1: DOUBLE STARS
Short Hills, NJ: Enslow, 1979, ISBN 0-89490-027-7 pbk; Guildford: Lutterworth Press, 1979, ISBN 0-7188-2433-4 pbk. 120p; diags. Bibl (2).

The first of a series of manuals designed primarily for the more experienced amateur, but which can be used profitably by the beginner. Each volume is complete in terms of its own subject coverage. 'This volume provides the observer with all the information necessary to guide him towards successful double star work'. Sets out in detail the methods of observation and the recording of results. Includes a catalogue of 3,876 measures of 522 double and multiple stars (1941–75).

254 Jones, Kenneth Glyn, [David A. Allen and Edmund S. Barker]
WEBB SOCIETY DEEP-SKY OBSERVER'S HANDBOOK, VOL. 2: PLANETARY AND GASEOUS NEBULAE
Short Hills, NJ: Enslow, 1979, ISBN 0-89490-028-5 pbk; Guildford: Lutterworth Press, 1979, ISBN 0-7188-2434-2 pbk. 149p; diags. Bibl (1).

This volume provides background information, a theoretical résumé, and detailed notes on observing method and practice. Includes annotated catalogues of 80 planetary, and 81 gaseous nebulae, with drawings.

255 Jones, Kenneth Glyn (ed.), [Edmund S. Barker]
WEBB SOCIETY DEEP-SKY OBSERVER'S HANDBOOK, VOL. 3: OPEN AND GLOBULAR CLUSTERS
Hillside, NJ: Enslow, 1980, ISBN 0-89490-034-X pbk; Guildford: Lutterworth Press, 1980, ISBN 0-7188-2468-7 pbk. 206p; diags. Bibl (1).

Following the introductory matter, this volume presents a catalogue of 176 open, and 63 globular star clusters, annotated, mostly with drawings, and some additional listings.

256 Jones, Kenneth Glyn (ed.), [Edmund S. Barker]
WEBB SOCIETY DEEP-SKY OBSERVER'S HANDBOOK, VOL. 4: GALAXIES
Hillside, NJ: Enslow, 1981, ISBN 0-89490-050-1 pbk; Guildford: Lutterworth Press, 1981, ISBN 0-7188-2527-6 pbk. 238p; diags. Bibl (2).

Following the introductory matter, this volume presents a catalogue of 275 selected objects, followed by drawings of 156 galaxy fields. The nine appendices include a list of additional objects.

257 Jones, Kenneth Glyn (ed.), [George S. Whiston]
WEBB SOCIETY DEEP-SKY OBSERVER'S HANDBOOK, VOL. 5: CLUSTERS OF GALAXIES
Hillside, NJ: Enslow, 1982, ISBN 0-89490-066-8 pbk; Guildford: Lutterworth Press, 1982, ISBN 0-7188-2552-7 pbk. 241p; diags. Bibl (2).

Like the preceding volumes this one is divided into two sections, the first being descriptive, providing background information and dealing with observational techniques. The catalogue section is divided into three parts: 1. the Virgo cluster, 2. Abell clusters, and 3. groups of galaxies.

258 JOURNAL OF THE AMERICAN ASSOCIATION OF VARIABLE STAR OBSERVERS
(AAVSO, 187 Concord Avenue, Cambridge, MA 02138, USA), 1972–, 2 issues/year.

A semi-professional journal which contains papers presented at meetings as well as submitted papers, some as abstracts only. Includes Association matters and book reviews. The AAVSO also publishes the *Bulletin*, usually once a year, which contains annual predictions of maxima and minima of long period variables; the irregular *Reports*, which contain computerized light curves of variables in the AAVSO observing programme, usually covering an interval of 1,000 days; *Alert Notices; Circulars;* and the *Solar Bulletin*, produced by the Solar Division. (*See also* 263).

259 Mallas, John H. and Evered Kreimer
THE MESSIER ALBUM: AN OBSERVER'S HANDBOOK
Cambridge, MA: Sky, 1978, ISBN 0-933346-04-2; Cambridge: Cambridge University Press, 1979, ISBN 0-521-23015-2. 248p; illus

(some col.), diags. Bibl (1).

Based on a series of articles in *Sky & Telescope*, 1967–70, this unusual book deals with the life of Charles Messier, as well as providing a guide for observers. It contains “Messier and his catalogue” by Owen Gingerich; a facsimile of Messier’s catalogue, reduced by two-thirds so that each page of the present work contains four pages of the original; the authors’ own Messier album, 110 objects with a photograph, drawing, map, NGC number, location, basic data, NGC description, and visual appearance; “Hints for beginning observers” by Gingerich. The last 31 pages present a selection of excellent colour photographs.

*260 Mullaney, James and Wallace McCall
THE FINEST DEEP SKY OBJECTS
Cambridge, MA: Sky. 32p; illus.

Two amateur astronomers describe 105 showpieces of the sky, including double and multiple stars, clusters, planetary nebulae and galaxies, which they have observed with a variety of instruments. Includes tables giving each object’s position and other data. (From *Scanning the skies* July 1985, p 14).

261 Newton, Jack and Philip Teece
THE CAMBRIDGE DEEP SKY ALBUM
Cambridge: Cambridge University Press, 1983, ISBN 0-521-25668-2. 126p; col. illus. Bibl (1), index (1).

A collection of 126 colour photographs, each covering a number of objects, but arranged by Messier number where possible. All the photographs are to the same scale to allow comparisons. Includes descriptive notes and astronomical data for each photograph. The introduction provides notes on observation and astrophotography.

262 Newton, Jack
DEEP SKY OBJECTS: A PHOTOGRAPHIC GUIDE FOR THE AMATEUR
Toronto: Gall Publications, 1977, ISBN 0-88904-081-8. 160p; illus, diags. Glos (1), bibl (1), index (1).

‘Prepared to help amateur astronomers locate and identify the Messier objects and some of the more interesting NGC numbers’. Includes full-page photographs of most of these objects, to roughly the same scale, with finder maps and brief descriptions. In some ways an early version of (261).

263 VARIABLE VIEWS
(Carolyn Hurless, 1745 Rice Avenue, Lima, OH 45805, USA),

monthly. Illus, diags. Q

This informal journal, intended primarily for members of the AAVSO, has been edited and produced single-handedly for over twenty years by Carolyn Hurless. It consists mainly of letters from readers and cuttings from other periodicals, mostly reproduced directly, with added editorial comments. Includes some specially written articles.

264 THE WEBB SOCIETY QUARTERLY JOURNAL
(S.J. Hynes, Secretary, 8 Cormorant Close, Sydney, Crewe, Cheshire, CW1 1LN, UK), quarterly. Illus, diags.

The purpose of the Webb Society is to promote the study of double stars and other deep sky objects among amateur astronomers. It is named after the Rev. T.W. Webb (1807–85), who wrote the important guide, *Celestial objects for common telescopes* (1859). The journal contains articles by amateur and professional astronomers on deep sky astronomy, covering observation, background information such as mythology, observing section reports, news, features for beginners, and book reviews.

A reprint of the sixth edition of Webb's book, 1917, edited and revised by Margaret W. Mayall, is still in print (New York: Dover, 1962, Vol. 1: *The Solar System*, ISBN 0-486-20917-2 pbk; Vol. 2: *The stars*, ISBN 0-486-20918-0 pbk).

Amateur Observing III — The Solar System

Reference/general

265 Chernan, Carl M.
THE HANDBOOK OF SOLAR FLARE MONITORING AND PROPAGATION FORECASTING
Blue Ridge Summit, PA: TAB Books, 1978, ISBN 0-8306-9984-8; 0-8306-1005-7 pbk. 194p; illus, diags. Index (1).

This book, intended for radio amateurs and experimenters, provides background information on wave propagation and the ionosphere, before going on to describe the methods of flare detection, and the construction of three solid state VLF receivers and recording interfaces. The appendices include a list of VLF stations.

266 Cherrington, Ernest H., jr.
EXPLORING THE MOON THROUGH BINOCULARS AND SMALL TELESCOPES
New York: Dover, 1984, ISBN 0-486-24491-1 pbk (corrected reprint of 1st ed. 1969). 229p; illus. Bibl (2), index (2). Q

This introduction for the beginning observer provides much background information, and a gazetteer of lunar features. But the main part of the book is a 28-day guide, with substantial notes, based on photographic charts showing the Moon at seven different phases, in labelled and unlabelled versions on facing pages. There are also 53 larger photographs showing sections of the lunar surface in greater detail, each with a smaller labelled version.

A similar volume, still sought after, is Henry Hatfield's *Amateur astronomer's photographic lunar atlas* (London: Lutterworth Press, 1968, ISBN 0-7188-1353-7, 116p).

*267 Hill, Richard
HANDBOOK FOR THE WHITE LIGHT OBSERVATION OF SOLAR PHENOMENA
Tucson, AZ: Association of Lunar and Planetary Observers, 1984, unbound. 47p.

A guide for the amateur astronomer interested in detailed sunspot observing, both visual and photographic. The individual sheets are held together by wire pins to ease substitution of amendments and addition of updates. Covers granulation, sunspot penumbras, the solar cycle, solar rotation, faculae, sunspot drawings, and the heliographic co-ordinate system. (From *S&T* Sept. 1985, p 226).

268 THE JOURNAL OF THE ASSOCIATION OF LUNAR AND PLANETARY OBSERVERS (THE STROLLING ASTRONOMER)
(ALPO, Box 3AZ, University Park, Las Cruces, NM 88003, USA), 1953–, quarterly, ISSN 0039-2502. Diags. Ads.

The ALPO, founded in 1947, is an informal, international group, whose purpose is to stimulate, co-ordinate, and generally promote the study of the bodies of the Solar System. The journal contains articles, reports of the observing sections, with interpretation, often including drawings and charts. Includes Association matters and book reviews. The sections of the ALPO are: Mercury, Venus, Mars, Jupiter, Saturn, Remote Planets, Solar, Lunar, Comets and Meteors (which produces a newsletter, *Tails and Trails*), and Minor Planets (which produces the quarterly *Minor Planets Bulletin*, Derald D. Nye, Route 7, Box 511, Tucson, AZ 85747, USA).

269 PLANETARY AND LUNAR CO-ORDINATES FOR THE YEARS 1984–2000
London: HMSO, 1983, ISBN 0-11-886917-5 pbk; Washington, DC: US Government Printing Office, 1983, pbk. 321p.

Prepared jointly by the Nautical Almanac Office at the Royal Greenwich Observatory and the US Naval Observatory, this is a continuation of a previous volume which covered the years 1980–84. It aims to provide low-precision data for planning purposes in advance of the annual publication of *The astronomical almanac* (207).

*270 Maag, Russell C., Jerry M. Sherlin and Rollin P. van Zandt
OBSERVE AND UNDERSTAND THE SUN: A PROGRAM FOR OBSERVING AND PHOTOGRAPHING THE SUN
Washington, DC: Astronomical League, 1977, pbk. 48p; diags.

A guide by experienced amateurs dealing with white-light observing, observing in the red light of the hydrogen-alpha line, solar photography, and solar radio astronomy. Includes details for constructing the necessary equipment. (From *S&T* Sept. 1977, p 221–2). A more detailed guide is W.M. Baxter's *The Sun and the amateur astronomer*, 2nd ed. (Newton Abbot: David & Charles, 1973, ISBN 0-7153-5629-1; New York: Drake, 1973, 165p).

*271 Mansfield, Roger
LOCAL PLANET VISIBILITY REPORT
Colorado Springs, CO: Astronomical Data Service, annual.

A computer printout custom-prepared for an observer's latitude (which should be supplied when ordering), which provides information at two-day intervals for Mercury and Venus, and at five-day intervals for Mars, Jupiter and Saturn. Includes each planet's altitude and azimuth at morning or evening, RA and declination, solar co-ordinates, distance from Earth and Sun, apparent diameter, solar elongation, illuminated portion of the disc, visual magnitude, and rising or setting time. (From *S&T* Feb. 1985, p 132).

272 Meeus, Jean
ASTRONOMICAL TABLES OF THE SUN, MOON AND PLANETS
Richmond, VA: Willmann-Bell, 1983, ISBN 0-943396-02-6 pbk. 384p. Bibl (1), index (2).

This book provides a wealth of detail which will interest professional as well as amateur astronomers, and will prove useful for historical research. The information provided often covers large time periods, e.g. oppositions of Mars 0-3010 AD, and sunspot activity 1749–1981, transits of Venus 1300–4100. Gives clear explanations throughout.

273 MINOR PLANET CIRCULARS
Cambridge, MA: Minor Planet Center, Smithsonian Astrophysical Observatory. About 1000/year.

These circulars, which since 1978 have included data on comets as well as minor planets, and also contain the official announcements of the new names of minor planets, are issued in monthly batches on the date of the full moon. Subscribers include most observatories and many advanced amateurs around the world.

274 Povenmire, Harold R.
GRAZE OBSERVER'S HANDBOOK (2nd ed.)
Indian Harbor Beach, FL: JSB Enterprises, 1979, ISBN 0-934086-00-1. 180p; illus, diags. Glos (1), bibl (1).

A grazing occultation is 'when a star or other astronomical body is covered up by the extreme north or south limb or edge of the Moon'. Such events are useful for making more accurate determinations of the Moon's position and shape, and in the detection of binary stars. The author gives a brief history of the subject, an introduction to the field, then goes on to discuss equipment, observing techniques, and the organization of expeditions. (First ed. 1975 only seen).

275 TONIGHT'S ASTEROIDS
(Dr. J.U. Gunter, 1411 North Mangum St., Durham, NC 27701, USA), bi-monthly. Diags.

This looseleaf publication (which is available for the cost of postage) usually contains abstracts of letters from subscribers, a regular column 'Minor Planet Remarks', fuller notes on featured asteroids and their discoverers, and often notes on new minor planet names. Also covers comets. One side of each sheet contains a tracking chart for a specific asteroid. Includes book reviews.

Meteors

*276 Kronk, Gary W.
CATALOG OF METEOR SHOWERS
Hillside, NJ: Enslow, 1983, ISBN 0-89490-072-2 pbk. 128p.

277 Mackenzie, Robert A.
BMS METEOR OBSERVER'S HANDBOOK, VOL. 1: NAKED EYE WORK (6th ed.)
Dover: British Meteor Society, 1982, pbk. 27p; diags. Glos (1). Q

This guide for serious amateurs covers observing techniques, calculating positions, deriving radiants, analysing observations, determining meteor trajectories in the atmosphere, and other relevant matters.

278 Mackenzie, Robert A.
BMS METEOR OBSERVER'S HANDBOOK, VOL. 2: TELESCOPIC METEOR WORK (6th ed.)
Dover: British Meteor Society, 1983, pbk. 10p; diags. Q

This volume covers optical equipment, observational techniques, the determination of radiant positions, magnitude distributions, sporadic and shower rate calculations, beginning and end height calculations. Assumes the reader has access to volume one. Further volumes (not seen) are: 3. *Photographic and spectrographic meteor work*, 4. *Collecting and studying micrometeorites*, 5. *Radio scatter meteor work*, and 6. *Meteor computation*.

279 Mackenzie, Robert A.
BMS RADIANT CATALOGUE (4th ed.)
Dover: British Meteor Society, 1981, pbk. 43p. Q

Following explanatory notes in English, French, Dutch, Italian, Spanish and Portuguese, is a list of 882 meteor streams in two tables, one giving data needed by the observer, the other giving reference data, orbital elements and parent bodies. Further tables give ephemerides of a few selected streams, notes on stationary radiants, and twin showers. Also sets out the rules for defining a new radiant (the apparent point of origin of a meteor stream).

*280 Mackenzie, Robert A.
SOLAR SYSTEM DEBRIS
Dover: British Meteor Society, 1980, pbk. 103p; diags. Q

281 METEOR NEWS
(Callahan Astronomical Society, Route 3, Box 424-99, Callahan, FL 32011, USA), quarterly, ISSN 0146-9959. About 12p; illus, diags. Q

Contains brief articles, news, lists of recent observations. Also covers meteorites.

282 METEOROS: JOURNAL OF THE BRITISH METEOR SOCIETY
(British Meteor Society, 26 Adrian St., Dover, Kent, CT17 9AT,

UK), 5 issues/year. About 12p; diags. Q

Despite its name the BMS has an international membership (50 per cent outside the UK according to its information sheet). The journal contains reports of research, observational results, comment, criticism, news, reviews and letters. Illustrations are usually line drawings, but occasionally include photographs.

283 Povenmire, Harold R.
FIREBALLS, METEORS AND METEORITES
Indian Harbor Beach, FL: JSB Enterprises, 1980, ISBN 0-934086-03-6. 215p; illus, diags. Bibl (2).

A straightforward readable guide for the beginning amateur observer, which provides much useful advice and a wealth of background information. Coverage includes meteor photography, tektites, comets and asteroids.

Comets

284 COMET NEWS SERVICE
(McDonnell Planetarium, 5100 Clayton Road, St. Louis, MO 63110, USA), quarterly. Diags. Q

A review of cometary news and research, with additional special bulletins for newly discovered bright comets.

285 Doherty, Paul
THE ARRIVAL OF HALLEY'S COMET
London: Aurum Press, 1985, ISBN 0-948149-09-4 pbk. 32p; illus. Q

A concise general guide (listed here before a copy was available), which covers topics such as comets in general, the returns of 1835 and 1910, and the relationship of meteors to comets. Includes a pull-out glow-in-the-dark viewing chart.

286 Edberg, Stephen J.
INTERNATIONAL HALLEY WATCH AMATEUR OBSERVER'S MANUAL FOR SCIENTIFIC COMET STUDIES, PART I: METHODS, PART II: EPHEMERIS AND STAR CHARTS
Hillside, NJ: Enslow, 1983, ISBN 0-89490-102-8 pbk; and Cambridge, MA: Sky, 1983. ISBN 0-933346-40-9 pbk. 92 + 91p; diags. Bibls (3). Q

A book for the advanced amateur. Part one provides instructions on the proper methods of generating meaningful scientific data on comets. The novice can learn general observing techniques while learning these methods. Part two contains an emphemeris and 19 detailed star charts for finding the comet and making useful observations of it. Includes report forms. The manual does not teach basic observing, telescope adjustment, or data reduction techniques. (Originally published in 2 vols. Pasadena, CA: Jet Propulsion Laboratory, 1983, JPL 83–16).

A newsletter for observers is the *International Halley Watch Amateur Observer's Bulletin* (Planetary Society, B12, 65 North Catalina, Pasadena, CA 91006, USA).

287 Hendrie, Michael
HOW TO SEE HALLEY'S COMET
London: Times Books, 1985, ISBN 0-7230-0273-8 pbk. 32p; diags.

A concise observational guide for the curious novice by the head of the British Astronomical Association's Comet Section, which includes some background information on comets in general and Halley's Comet in particular.

288 INTERNATIONAL COMET QUARTERLY
(Daniel Green, Smithsonian Astrophysical Observatory, 60 Garden St., Cambridge, MA 02138, USA), 1979–, quarterly, ISSN 0736-6922. Illus (on cover). Q

'A non-profit journal devoted to news and observation of comets' which acts as a link between amateur and professional astronomers. Includes some longer articles on topics of current interest. (This is a successor to *The Comet*).

*289 Kronk, Gary W.
COMETS: A DESCRIPTIVE CATALOG
Hillside, NJ: Enslow, 1984, ISBN 0-89490-071-4 pbk. 331p. Bibl, index.

The result of several years' work by an amateur astronomer, this catalogue divides the descriptive material into sections for short- and long-period comets. (From *S&T* May 1985, p 419).

*290 Machholz, Don
A DECADE OF COMETS
San Jose, CA: D. Machholz, 1985, pbk. 111p.

The author, an amateur astronomer, compares comet discoveries by amateur and professional astronomers between the start of 1975 and the end of 1984, a period in which amateurs discovered 33 comets, 38 per cent of all those found. (From *Astronomy* Sept. 1985, p 45).

291 Mackenzie, Robert A.
HANDBOOK OF COMETARY STATISTICS
Dover: British Meteor Society, 1982, pbk. 13p.

Cometary orbital data in tabular form, covering comet groups, distances of nodes from the Sun, and distributions.

*292 Mansfield, Roger
COMET HALLEY VISIBILITY REPORT 1985–86
Colorado Springs, CO: Astronomical Data Service, 1985. Bibl.

A computer printout specially prepared for the observer's latitude, which provides information on the position of the comet and its visual magnitude for the period August 1985 to August 1986. The *User's guide* gives background information, including observing notes. (Publisher's information). (Write to: 3922 Leisure Lane, Colorado Springs, CO 80917, USA).

*293 Marsden, Brian G.
CATALOG OF COMETARY ORBITS
Hillside, NJ: Enslow, 1983, ISBN 0-89490-095-1 pbk. 96p.

Lists orbital elements for 1,109 cometary apparitions, all those observed up to the end of May 1982. Other sections provide names, appearances of periodic comets, non-gravitational parameters, references, and listings of the comets sorted by orbital class. Prepared from computer copy. (From *S&T* Oct. 1983, p 321).

294 Muirden, James
OBSERVER'S GUIDE TO HALLEY'S COMET
New York: Arco, 1985, ISBN 0-668-06377-7 pbk; London: George Philip, 1985, ISBN 0-540-01095-2 pbk. 75p; illus, diags.

This clear, concise guide includes finder charts, constellation maps and a monthly log from December 1985 to June 1986, as well as background information on comets in general and Halley's Comet in particular.

*295 THE RETURN OF HALLEY'S COMET
San Francisco: Astronomical Society of the Pacific, 1985, pbk. 36p; illus, diags. Q

Tells the novice when and where to look for Comet Halley during the 1985–86 return. The charts are adapted from Donald K. Yeomans' *The Comet Halley handbook* (298). Includes five articles on comets from *Mercury* (139). (From *S&T* Aug. 1985, p 133, and *Astronomy* Oct. 1985, p 86–7).

*296 Stasiuk, Garry and Dwight Gruber
THE COMET HANDBOOK
Portland, OR: Stasiuk Enterprises, 1984, ISBN 0-932421-00-8 pbk. 32p.

A book about Halley's Comet with emphasis on observing. Includes an 'observing calendar' from October 1985 to May 1986 giving the Moon's phases, and suggesting when and where to look, as well as a number of charts to aid the observer. There is a section on comets in history. The final five pages are devoted to observing techniques. (From *S&T* Aug. 1985, p 131).

297 Tattersfield, D.
ORBITS FOR AMATEURS: WITH A MICROCOMPUTER
Cheltenham: Stanley Thornes, 1984, ISBN 0-85950-146-9; New York: Halsted Press, 1984, 0-470-27502-2. 171p. Bibl (1).

An introduction to the calculations associated with cometary orbits which have in the past been considered too complex for amateurs. Provides step by step instructions for solving the four main problems: the construction of an ephemeris, determination of orbital elements, perturbations, and differential corrections. Includes programs in BASIC for the Spectrum ZX 81 computer and the Sharp PC 1211 or Tandy TC-1 programmable calculators. (*See S&T* April 1985, p 322–3 for some problems likely to be encountered in using this book).

298 Yeomans, Donald K.
THE COMET HALLEY HANDBOOK: AN OBSERVER'S GUIDE (2nd ed.)
Pasadena, CA: Jet Propulsion Laboratory, 1983, pbk. 44p; diags. Bibl (1). Q

This guide for the serious amateur and professional astronomer provides notes on the orbit of Comet Halley, its expected physical behaviour in 1985–86, observing conditions 1985–86, aided by numerous diagrams. Appendices give historical, physical and orbital data, and include 20 pages of ephemeris data 1981–87. (1st ed. 1981 only seen).

Laboratory Manuals

Most of the following books are intended for use in a formal teaching environment, and may often be used in conjunction with the textbooks listed earlier (57–113), but generally they will also be suitable for private study. (*See also* 89 and 102).

299 Christiansen, Wayne A., Ronald H. Kaitchuk and Michelle D. Kaitchuk
INVESTIGATIONS IN OBSERVATIONAL ASTRONOMY (2nd ed.)
Geneva, IL: Paladin House, 1978, ISBN 0-88252-054-7 spiral-bound. 165p; illus, diags. Q

Twenty exercises for an introductory college-level course in introductory observational astronomy. Some are indoor projects, while others require small telescopes or access to a planetarium. Uses some algebra.

300 Culver, Roger B.
AN INTRODUCTION TO EXPERIMENTAL ASTRONOMY: AN OBSERVATIONAL WORKBOOK
New York: Freeman, 1984, ISBN 0-7167-1495-7 pbk. 196p; illus, diags. Q

Intended to accompany elementary astronomy courses, this book, reproduced from typescript, presents 22 experiments which involve algebraic calculations, e.g. the distance between Earth and Sun, the mass of Jupiter, properties of the Milky Way, observation of a quasar. (Based on a preliminary version which appeared in 1974).

301 Greenleaf, Peter
EXPERIMENTS IN SPACE SCIENCE
New York: Arco, 1981, ISBN 0-668-04812-3 pbk. 166p; illus, diags. Glos (2), index (1).

Based on an earlier version entitled *Let's explore outer space* (1969), this book is intended primarily for younger students, but the content is less juvenile than the title of the first version suggests. The cover proclaims, 'Learn how to solve research problems in space science', which promises rather more than the book delivers: a set of simple astronomical experiments, with some problems involving rocketry. Includes review questions, with answers.

302 Hoff, Darrel B., Linda J. Kelsey and John S. Neff
ACTIVITIES IN ASTRONOMY
Dubuque, IA: Kendall/Hunt, 1978, ISBN 0-8403-1849-9 spiral-bound. 153p; illus, diags. Bibl (x). Q

Attempts 'to meet the needs of one-semester introductory astronomy lab courses'. Several of the 25 exercises, with choices for both outdoor observing and classroom work, have been shortened or simplified from (304, 1st ed.). Appendices include guidance on telescope alignment, dark-room techniques, provide data sheets and graph paper, and a Messier catalogue.

*303 Kafatos, Minas
ASTRONOMY LABORATORY MANUAL
Dubuque, IA: Kendall/Hunt, 1984, ISBN 0-8403-3405-2 spiral-bound. 83p.

A manual for an introductory course. All the exercises take place indoors, and little specifically astronomical equipment is required. (From *S&T* Feb. 1985. p 132).

304 Kelsey, Linda, Darrel Hoff and John Neff
ASTRONOMY: ACTIVITIES AND EXPERIMENTS (2nd ed.)
Dubuque, IA: Kendall/Hunt, 1983, ISBN 0-8403-2958-X pbk. 216p; illus, diags. Q

Intended as an aid to teaching astronomy at an introductory level, this manual describes 43 exercises, and includes discussion questions. The second edition includes an appendix of computer programs in Applesoft BASIC. Additional material, including a selection of reprints from *Sky & Telescope*, is provided in a separate folder.

305 Knox, Richard
EXPERIMENTS IN ASTRONOMY FOR AMATEURS
Shaldon, Devon: Keith Reid, 1975, ISBN 0-904094-15-4; New York: St. Martin's, 1976, ISBN 0-312-27685-0; 1979, ISBN 0-312-27686-9 pbk. 206p; diags. Bibl (1), index (2).

A guide to practical astronomy without the need for a telescope. Intended for the interested beginner rather than the student. Requires only elementary mathematical knowledge. Covers the whole field of astronomy, though far less rigorously than Tattersfield (312). Exercises follow each chapter.

306 Lacy, Claud H.
ASTRONOMY LABORATORY EXERCISES
Dubuque, IA: Kendall/Hunt, 1981, ISBN 0-8403-2564-9. 65p. Q

A set of twelve basic exercises covering topics such as the bodies of the Solar System, pulsars, and the expansion of the universe. Includes reprints from *Sky & Telescope*.

*307 Payne, George
NAKED EYE ASTRONOMY
Geneva, IL: Paladin House, 1980, ISBN 0-88252-102-0 pbk.

*308 Payne, George
SOLAR SYSTEM ASTRONOMY
Geneva, IL: Paladin House, 1980, ISBN 0-88252-103-9 pbk.

309 Robbins, Robert R. and Mary Kay Hemenway
MODERN ASTRONOMY: AN ACTIVITIES APPROACH
Austin, TX: University of Texas Press, 1982, ISBN 0-292-75064-1 pbk. 272 + 2p; illus, diags. Bibl (x), index (2). Q

A practical approach to learning astronomy for students and other interested parties. Begins with the basic activities of measurement, and mapping the night sky; once these are mastered the rest can be tackled in any order. They include planetary motions, the Sun, properties of lenses and mirrors, using a small telescope, an introduction to spectroscopy, an introduction to computers and BASIC, and photography. Includes a cross staff and quadrant, and a spectrometer scale on card to cut out, but some of the equipment used is rather advanced for the reader studying at home.

*310 Safko, L.
LABORATORY EXERCISES IN ASTRONOMY (2nd ed.)
Dubuque, IA: Kendall/Hunt, 1978. 380p.

*311 Schroeder, Leon
INTRODUCTION TO ASTRONOMY: AN INDEPENDENT STUDY MANUAL
Geneva, IL: Paladin House, 1978, ISBN 0-88252-077-6 spiral-bound. 121p.

Though intended for independent study students this book adapts easily to lecture courses as an assignment manual. The 25 units cover the Solar System, telescopes, varieties of stars, spectral classes, galaxies and cosmology. (From *S&T* May 1979, p 478–9).

312 Tattersfield, D.
PROJECTS AND DEMONSTRATIONS IN ASTRONOMY
Cheltenham: Stanley Thornes, 1979, ISBN 0-85950-087-X. 331p; diags. Index (2).

Intended for anyone seriously studying astronomy below the university level, this book provides a practical approach to learning, without the need for a telescope. Includes 61 projects, mostly to be carried out using second-hand data. Each project is preceded by the necessary theoretical background, which requires knowledge of elementary algebra. The projects range over the whole field of astronomy. Includes solutions to the problems set.

313 Waxman, Jerry
A WORKBOOK FOR ASTRONOMY
Cambridge: Cambridge University Press, 1984, ISBN 0-521-25312-8 comb-bound. 356p; illus, diags. Glos (2), index (2).

A comprehensive set of exercises at varying levels, mostly self-contained or requiring equipment which can be constructed from plans provided; a few are for students with access to a planetarium. Acts as a supplement to the factual material in a textbook: some of the exercises are brief enough to accompany a class lecture, while others are suitable for a three-hour laboratory period. Uses some algebra.

Equipment and Techniques

Telescopes — general/selection and use

314 Bell, Louis
THE TELESCOPE
New York: Dover, 1981, 0-486-24151-3 pbk. 287p; illus, diags. Bibl (1), index (2).

This reprint includes a new introduction by Jay M. Pasachoff, who comments that the book is 'surprisingly timely and up to date', and summarizes developments since the book was originally published in 1922, noting the development of the Schmidt wide-field camera which overcame the problem of spherical aberration that much concerned the author. The book provides a brief history of the telescope, but is basically a manual for the amateur, covering optics, construction, mechanics, testing, maintenance, and use. Does not cover photography. The brief bibliography is a new one.

315 Brown, Sam
ALL ABOUT TELESCOPES (4th ed.)
Barrington, NJ: Edmund Scientific, 1979. 192p; diags. Q

If the title is not literally true, this tightly packed compendium, with hundreds of small line drawings in semi-cartoon style, certainly gives the impression of completeness. It is a guide for beginners to observing, photography, mirror grinding and testing, telescope construction, mounts, collimation and adjustment, and telescope optics.

316 Levenson, Jordan
HOW TO BUY AND UNDERSTAND REFRACTING TELESCOPES
Los Angeles: Levenson Press, 1981, ISBN 0-914442-09-0 pbk. 97p; diags. Index (1).

A guide for the serious beginner which covers eyepieces,

magnification limits, resolution, tubes and mounts. The author explains how the materials and manufacturing process of telescopic optics determine the quality and price of a telescope.

*317 Lovi, George, (Saul Lapidus, ed.)
THE TELESCOPE CATALOG OF CATALOGS
New York: D. McKay, 1979, pbk. 131p; diags.

The first part of this guide for amateurs explains, with many diagrams, the various types of telescope, eyepiece and mounting. The second part reproduces 55 pages of telescope advertisements (including specifications) from leading suppliers. (From *S&T* Nov. 1979, p 459).

*318 Mansfield, Roger
SPECIAL REPORT TO CELESTRON TELESCOPE OWNERS
Colorado Springs, CO: Astronomical Data Service, 1980, pbk. 24p; diags.

A mimeographed pamphlet which provides aid in the setting up and polar alignment of Celestron telescopes. Includes general information on the astronomical co-ordinate system and principles of the equatorial mounting. The material is mostly applicable to any fork-mounted, compact telescope that has a wedge-type latitude adjustment. (From *S&T* Oct. 1980, p 321).

*319 Moore, Patrick
HOW TO MAKE THE MOST OF YOUR TELESCOPE
Harlow: Longman, 1985, ISBN 0-582-23535-9. 96p; illus, diags. Index.

A basic guide for those with access to a small, low-powered telescope. Coverage includes care of the telescope, observing various objects, and careers in astronomy. (Publisher's information).

*320 Muirden, James
HOW TO USE AN ASTRONOMICAL TELESCOPE
New York: Simon & Schuster (Linden Press), 1985, ISBN 0-671-47744-7. 480p.

*321 Paul, Henry E., (revised by Greg Stone)
BINOCULARS AND ALL PURPOSE TELESCOPES (2nd ed.)
New York: Amphoto, 1980, ISBN 0-8174-3558-1; 0-8174-3559-X. 96p.

Advice for the beginner on choosing and buying observing instruments, devoted largely to binoculars, with only ten pages on telescopes, and briefer coverage of long-range photography. (From *Observatory* Feb. 1982, p 14).

322 Paul, Henry E.
TELESCOPES FOR SKYGAZING (3rd ed.)
New York: Amphoto, 1976, ISBN 0-8174-2408-3. 160 + 4p; illus (some col.), diags. Bibl (1), index (1).

This beginner's guide starts with a survey of objects to observe, before going on to discuss the different types of telescope, the advantages of reflectors over refractors, accessories, 'build, buy, or assemble?', testing and care, using binoculars, and including an introduction to astrophotography. (2nd ed. only listed by Seal (24)).

323 Robinson, J. Hedley
USING THE TELESCOPE: A HANDBOOK FOR ASTRONOMERS
Newton Abbot: David & Charles, 1978, ISBN 0-7153-7592-X; New York: Halsted Press, 1979, ISBN 0-470-26514-0. 112p; illus, diags. Bibl (1), index (1).

A practical guide for the beginning observer, which covers instruments: choosing, mounting, housing, eyepieces, accessories, maintenance; as well as photography; time, the clock and position; and observing the major bodies of the Solar System.

324 Taylor, H. Dennis
THE ADJUSTMENT AND TESTING OF TELESCOPE OBJECTIVES (5th ed.)
Bristol: Adam Hilger, 1983, ISBN 0-85274-756-X pbk. 103p; diags.

This guide for owners of refracting telescopes was first published in 1891 and has changed little through its five editions, attesting the usefulness of its advice.

325 Traister, Robert J. and Susan E. Harris
ASTRONOMY & TELESCOPES: A BEGINNER'S HANDBOOK
Blue Ridge Summit, PA: TAB Books, 1983, ISBN 0-8306-1419-2 pbk. 200p; illus (some col.), diags. Glos (2), index (2).

Though this book provides a brief introduction to astronomy, it is largely a guide to various commercial telescopes, and to the construction, maintenance and use of telescopes.

*326 Trueblood, Mark and Russell M. Genet
MICROCOMPUTER CONTROL OF TELESCOPES
Richmond, VA: Willmann-Bell, 1984, ISBN 0-943396-05-0 pbk. 300p.

Reflecting the new possibilities for telescope control, this book deals with microcomputer control of small- and medium-sized telescopes. Three parts cover: 1. theoretical and practical considerations, 2. hardware (motors, resolvers, etc.), 3. detailed descriptions of several working systems of interest to amateurs and small observatories. (Publisher's information).

Telescopes and observatories — construction

*327 [Berry, Richard (ed.)]
HOW TO BUILD A DOBSONIAN TELESCOPE (2nd ed.)
Milwaukee, WI: AstroMedia, 1983, pbk. 52p; illus, diags. Q

A collection of articles reprinted from *Telescope Making* (337), giving instructions for the construction of a ten-inch Dobsonian, with information and photographs on 12-, 16- and 20-inch telescopes. The Dobsonian is a particularly stable type of telescope, which reduces the amount of vibration thus enabling the use of higher-power eyepieces. (From *Spaceflight* Dec. 1981, p 90, and publisher's information).

328 [Berry, Richard (ed.)]
HOW TO BUILD YOUR OWN OBSERVATORY
Milwaukee, WI: AstroMedia, 1981, pbk. 42p; illus, diags. Q

A collection of ten short articles reprinted from *Telescope Making* (337), describing observatories made by readers.

*329 Howard, Neale E.
STANDARD HANDBOOK FOR TELESCOPE MAKING (2nd ed.)
New York: Harper & Row, 1984, ISBN 0-06-181394-X. 356p.

Step by step instructions for building an eight-inch Newtonian reflector. While the bulk of the text devoted to mirror making remains unchanged from the first edition of 1959 (UK 1962, omitting the word *Standard*), there is new material on mountings (including Dobsonians), backyard observatories, celestial photography and observing. An appendix lists suppliers of telescope making parts. (From *S&T* March 1984, p 238).

330 Ingalls, Albert (ed.)
AMATEUR TELESCOPE MAKING, BOOK ONE (4th ed.)
New York: Scientific American, 1974 (reprint of 4th ed. 1945). 510p; illus, diags. Index (3).

The classic manual for the serious telescope maker. This volume is intended for beginners as well as the more advanced amateur. Because the contributions were built up over several editions information tends to be scattered throughout the book, making use of the index essential. Not listed by Seal (24).

331 Ingalls, Albert (ed.)
AMATEUR TELESCOPE MAKING ADVANCED, BOOK TWO
New York: Scientific American, 1970 (reprint of rev. impression, 1946, of 1st ed.). 650p; illus, diags. Index (1).

Subtitled 'a collection of contributions to amateur precision optics by numerous authorities', this volume is divided into two parts: 1. construction of optical instruments (over 500 pages), and 2. practical aspects of observing.

332 Ingalls, Albert (ed.)
AMATEUR TELESCOPE MAKING, BOOK THREE
New York: Scientific American, 1969 (reprint of 1st ed. 1953). 646p; illus, diags. Index (2).

Subtitled 'contributions to amateur precision optics by advanced amateurs and professionals', this volume is a miscellany essentially for the enthusiastic amateur, but also for anyone interested in optics. Includes brief biographies of the contributors.

333 Muirden, James
BEGINNER'S GUIDE TO ASTRONOMICAL TELESCOPE MAKING
London: Pelham Books, 1978, ISBN 0-7207-0822-2 (reprint of 1st ed. 1975). 201 + 9p; illus, diags. Bibl (1), index (1).

After introducing telescopes in general, the author gives a step by step guide to the construction of a reflecting telescope. Covers planning; the mirror: grinding and smoothing, polishing and testing, figuring; tubes and mounts; optical alignment and accessories; designing and making an object glass; larger reflectors and advanced telescopes. Also describes how to make a simple refractor.

*334 Sickels, R. Mansfield
RADIO ASTRONOMY HANDBOOK
Fort Pierce, FL: Radio Astronomy Systems, 1984. 191p (in looseleaf binder).

Technical information for the beginning home radio astronomer who has some background in radio. Reproduced from typescript, it is similar to the author's *Radio astronomer's notebook* (with Jeff Lichtman, 1975), and *Amateur radio astronomer's notebook, vol. 2*, (1977). (From *S&T* Jan. 1985, p 36).

A guide for the radio amateur and electronic experimenter which assumes mastery of the basic skills of circuit construction is Dave Heiserman's *Radio astronomy for the amateur* (Blue Ridge Summit, PA: TAB Books, 1975, ISBN 0-8306-4714-7 pbk, 251p).

335 Spry, Reg
MAKE YOUR OWN TELESCOPE: FROM EVERYDAY MATERIALS
London: Sidgwick & Jackson, 1978, ISBN 0-283-98507-0; 0-283-98475-0 pbk. 48p; illus, diags.

Describes how to construct a telescope from materials such as planks, Meccano parts and coffee-jar tops, at a cost (in 1975) of £25.00.

336 Swenson, George W.
AN AMATEUR RADIO TELESCOPE
Tucson, AZ: Pachart, 1980, ISBN 0-912918-06-3 pbk. 58p; diags.

Describes a project carried out by the author's students, to design and build a small radio telescope of the phase-switching interferometer type. Intended to show that radio astronomy can be carried out by the amateur with knowledge of electronics. Includes circuit diagrams, and gives a brief introduction to cosmic radio waves, as well as noting objects to observe. This book, which first appeared as a series of articles in *Sky & Telescope* (May–Oct. 1978), has no contents page or index.

337 TELESCOPE MAKING
(AstroMedia Corp., 625 East St. Paul Avenue, PO Box 92788, Milwaukee, WI 53202, USA), quarterly, ISSN 0190-5570. About 48p; illus, diags. Ads. Q

'The magazine *for, by* and *about* telescope makers' covers instrumentation and observatories, but not objects to observe. Includes articles, reports on events for telescope makers, classified advertisements and correspondence. (*See also* 327 and 328).

338 Texereau, Jean, (Allen Strickler, trans.)
HOW TO MAKE A TELESCOPE (2nd ed.)
Richmond, VA: Willmann-Bell, 1984, ISBN 0-943396-04-2. (reprint of 2nd French ed. 1961). 424p; illus, diags. Bibls (3), indexes (2).

This highly regarded work begins with a discussion of basic principles, followed by a detailed guide to the construction of a Newtonian telescope. Then, using the experience gained, the reader is shown how to make a Cassegrainian telescope. Also covers making a telescope window, eyepieces (selection and types), the equatorial mounting, accessories, the adjustment of mirrors and mountings, and atmospheric turbulence. New appendices (1984) include a list of suppliers, computer programs, articles on telescope making in *Sky & Telescope, Telescope Making* and *Scientific American*, and a short professional biography of the author.

*339 Thompson, Allyn J.
MAKING YOUR OWN TELESCOPE (rev. ed.)
Cambridge, MA: Sky, 1980, ISBN 0-933346-12-3 pbk (reprint of rev. ed. 1947). 211p; illus, diags. Bibl, index.

This manual tells how to construct a six-inch reflecting telescope, though the procedures described can be applied equally well to an eight-inch or even a ten-inch reflector later on. (From *Scanning the skies* July 1985, p 30).

340 Worvill, Roy
TELESCOPE MAKING FOR BEGINNERS
London: Kahn & Averill, 1984, ISBN 0-900707-80-1 pbk (reprint of rev. impression, 1976, of 1st ed. 1974). 79p; diags. Bibl (2).

A brief guide to the construction of a small refractor and a reflecting telescope, with notes on observing various objects, using binoculars, a Sun clock and star clock.

Astrophotography

*341 ASTROPHOTOGRAPHY BASICS
Rochester, NY: Eastman Kodak, 1978, pbk (Kodak Publication AC-48).

A booklet on the fundamentals of astronomical photography.

Not seen, but probably still in existence, is the periodical *The Astrograph*, 1973–? (last known address: Box 2283, Arlington, VA 22202, USA).

*342 Barbera, R.F., C.F. Capen, B. Carvalho, snr. and R.A. Steeg
ASTRO FILTERS FOR OBSERVATION AND ASTROPHOTOGRAPHY (3rd ed.)
Oakland, CA: Optica, 1980, pbk. 102p.

This book provides background information on the nature of light and how filters work, along with practical matters, such as matching film and filter characteristics, focusing through filters, and using filters in combination with one another. Gives advice on selection of filters for different applications. Basically unchanged from the 2nd ed. (1973), this edition excludes nebula filters. (From *S&T* May 1981, p 440).

343 Cocozza, Joseph A.
ASTROPHOTOGRAPHY NEAR CITY LIGHTS
[Westville, NJ: J.A. Cocozza], 1977, pbk (boxed). 48p; illus, diags.
A collection of photographs, with text providing a brief guide to celestial photography. Topics covered include eliminating background fogging, choice of equipment, filter and film combinations, guiding systems, print enhancement, lunar and planetary photography.

*344 Covington, Michael
BASIC ASTROPHOTOGRAPHY
Cambridge: Cambridge University Press, 1985, ISBN 0-521-25391-8.
How to take pictures of stars, galaxies, planets, the Moon, the Sun, comets, meteors and eclipses, using equipment readily available to the amateur astronomer. (Publisher's information).

Also announced is, R.T. Little, *Astrophotography* (New York: Collier Macmillan, 1985, ISBN 0-02-948980-6).

*345 Gordon, Barry
ASTROPHOTOGRAPHY (2nd ed.)
Richmond, VA: Willmann-Bell, 1985, ISBN 0-943396-07-7.
A guide for beginners and more advanced enthusiasts which requires only elementary arithmetic. Presents the 'fx system', an approach to determining the correct exposure which assigns numerical values to the elements involved, requiring only simple addition to arrive at the correct fx for a scene. Covers the photography of various subjects, with a fixed camera, and long exposures with a driven equatorial mounting; optical systems; long focus photography; and includes a descriptive chapter on exotic equipment, such as Schmidt cameras, cold cameras and hypersensitized film. (From review of 1st ed. 1983, *S&T* Dec. 1983, p 518–20).

*346 Henderson, Robert V.
EXPOSURE GUIDE FOR ASTROPHOTOGRAPHY
Lynn, MA: Astronomics, 1977, pbk. 53p.
Suggestions for photographing the Sun, Moon, planets and other celestial objects are covered in 22 pages of text, followed by 31 pages of exposure data. Reproduced from typescript. (From *S&T* Sept. 1977, p 230).

*347 Lowenthal, James
THE HIDDEN SUN: SOLAR ECLIPSES AND ASTROPHOTOGRAPHY
New York: Avon, 1984, ISBN 0-380-86959-4 pbk, 107 + 16p; illus (some col.). Bibl.

A basic step by step guide to sky photography for beginners. Describes how to take pictures of eclipses, the Moon, stars, planets and deep-sky objects. (From *Scanning the skies* July 1985, p 19).

Photometry

*348 Genet, Russell M. (and others) (ed.)
NEAR IR PHOTOMETRY
Fairborn, OH: Fairborn Observatory, 1984, ISBN 0-911351-07-8 pbk. 220p.

349 Genet, Russell M.
SOLAR SYSTEM PHOTOMETRY HANDBOOK
Richmond, VA: Willmann-Bell, 1983, ISBN 0-943396-03-4 pbk. 189p; illus, diags. Bibl (x), index (1).

Aims to provide a concise introduction for advanced amateurs and smaller college observatories. The first section covers low-speed photometry: of asteroids, planets and satellites, comets, the Moon and the Sun, and low-speed equipment. The second section deals with high-speed photometry: occultations by planets and satellites, asteroid occultations, lunar occultations, and includes a portable high-speed photometer project.

350 Ghedini, Silvano
SOFTWARE FOR PHOTOMETRIC ASTRONOMY
Richmond, VA: Willmann-Bell, 1982, ISBN 0-943396-00-X. 221p; diags. Bibl (1), index (1).

This book is designed to meet the increasing need for programming required by amateurs with computers who wish to enter the field of photometric astronomy of variable stars, and other stellar objects. Describes a selection of algorithms, mostly dealing with the reduction of photoelectric observations to standard colour systems, and the analytical methods for variable star studies. Includes program listings in the Hewlett-Packard extended BASIC language.

351 Hall, Douglas S. and Russell M. Genet
PHOTOELECTRIC PHOTOMETRY OF VARIABLE STARS: A PRACTICAL GUIDE FOR THE SMALLER OBSERVATORY
Fairborn, OH: International Amateur-Professional Photoelectric Photometry (IAPPP), 1982. ring-bound. 200p; illus, diags. Bibl (2).

This book deals with the practical details of observatories, telescopes, photoelectric equipment, observing techniques, and the planning of observing programmes. Includes notes on the history of photometry, over four pages of abbreviations and acronyms, and a list of addresses, but no index.

352 Henden, Arne A. and Ronald H. Kaitchuk
ASTRONOMICAL PHOTOMETRY
New York: Van Nostrand, 1982, ISBN 0-442-23647-6. 392p; illus, diags. Bibl (x), index (2).

A self-contained introduction for both professionals and serious amateurs. The authors state that an amateur with little or no college background can use the book, but the mathematical content may prove daunting. Includes sections on observational techniques, the construction of equipment, computer programs for data reduction, and describes the applications of photometry, with several projects and tables of good objects to study.

353 Wolpert, Robert C. and Russell M. Genet (ed.)
ADVANCES IN PHOTOELECTRIC PHOTOMETRY, VOL. 1
Fairborn, OH: Fairborn Observatory, 1983, ISBN 0-911351-01-9 pbk. 237p; illus, diags.

Intended for the advanced amateur and the professional astronomer, this volume covers details of equipment design and construction, providing a guide for those wishing to enter the field of photoelectric photometry.

*354 Wolpert, Robert C. and Russell M. Genet (ed.)
ADVANCES IN PHOTOELECTRIC PHOTOMETRY, VOL. 2
Fairborn, OH: Fairborn Observatory, 1984, ISBN 0-911351-05-1 pbk. 185p.

The emphasis in this volume is less on basic equipment design and more on aspects of observing techniques. (From *JBAA* April 1985, p 136).

Calculators and computers

355 Burgess, Eric
CELESTIAL BASIC: ASTRONOMY ON YOUR COMPUTER
Berkeley, CA: Sybex, 1982, ISBN 0-89588-087-3 pbk. 300p; illus, diags. Bibl (1), index (2).

Designed for 'amateur and armchair astronomers and students' this book contains 23 general astronomical programs (and three alternatives) written for the Apple II computer in BASIC, but most are usable on a wide range of similar machines without major changes. The programs can be upgraded for greater precision. (*See also* 356).

356 Burgess, Eric and Howard J.
MORE USES FOR YOUR TIMEX/SINCLAIR 1000: ASTRONOMY ON YOUR COMPUTER
Berkeley, CA: Sybex, 1983, ISBN 0-89588-112-8 pbk. 153p; diags. Bibl (1), index (1).

A collection of 20 ready to run programs for the Sinclair ZX 81, the Timex/Sinclair 1000 and the Sinclair Spectrum, written in Sinclair BASIC. Most were first published in (355), but have been rewritten to make them accessible to a greater number of home computer owners. Unlike the earlier book this one does not provide background information on the topics covered. The programs are divided into four parts: time, the Moon, the planets, and general and tutorial.

357 Duffett-Smith, Peter
PRACTICAL ASTRONOMY WITH YOUR CALCULATOR (2nd ed.)
Cambridge: Cambridge University Press, 1981, ISBN 0-521-24059-X; 0-521-28411-2 pbk (both comb-bound). 188p; diags. Glos (2), index (2).

A selection of 71 calculations which should be within the capabilities of the keen beginner. The worked examples cover time, co-ordinate systems, the Sun, the planets, comets, binary stars, the Moon, and eclipses. Uses epoch 1980.0 as the basis for orbital calculations. Intended for any type of calculator, from the basic scientific to the programmable.

Announced for 1985 is the author's *Astronomy with your personal computer* (Cambridge University Press, 1985, ISBN 0-521-26620-3; 0-521-31976-5 spiral-bound).

358 Gavin, Maurice
ZX SPECTRUM ASTRONOMY: DISCOVER THE HEAVENS ON YOUR COMPUTER
London: Scot Press (Sunshine Books), 1984, ISBN 0-946408-24-6 pbk. 225p; diags.

Aimed at Spectrum owners who wish to extend their interest to astronomy, this book includes a variety of programs, written on a 48K Spectrum, all but one of which will run on the 16K version. The programs emphasize the visual aspects of astronomy rather than the manipulation of large amounts of numerical data. They are accompanied by notes and explanations, and cover time, co-ordinate systems, the Moon, satellites, the Solar System, star systems and starcharts.

359 Genet, Russell M. (ed.)
MICROCOMPUTERS IN ASTRONOMY
Fairborn, OH: Fairborn Observatory, 1983, ISBN 0-911351-03-5 pbk. 253p; illus, diags. Bibl (x).

A collection of 30 papers from a symposium held at the Fairborn Observatory, July 1983, divided into seven sections: large telescope control; small telescope control; slow photometer control; fast photometer control; computers, S-100 and Forth (including selection of a computer); analysis in photometry; and other microcomputer analysis.

*360 Genet, Russell M. and Karen A. (ed.)
MICROCOMPUTERS IN ASTRONOMY II
Fairborn, OH: Fairborn Observatory, 1984, ISBN 0-911351-06-X pbk. 224p.

361 Jones, Aubrey
MATHEMATICAL ASTRONOMY WITH A POCKET CALCULATOR
Newton Abbot: David & Charles, 1978, ISBN 0-7153-7675-6; New York: Halsted Press, 1978, ISBN 0-470-26552-3. 254p; diags.

Intended for the advanced amateur, to provide formulae, precepts and calculator programs for a wide range of problems: time, precession; positions of the Sun, Moon, and planets; comet and asteroid ephemerides; lunar eclipses; double stars; and ordinary and grazing occultation predictions. Some sections are for simple algebraic-logic calculators, others are for programmable HP-25 and HP-67 calculators.

362 Klein, Fred
POCKET COMPUTER PROGRAMS FOR ASTRONOMERS: A COLLECTION OF 13 PROGRAMS FOR OBSERVERS, ASTRO-PHOTOGRAPHERS AND DOBSONIANS
Los Altos, CA: Klein Publications, 1983, ISBN 0-913051-01-2 pbk. 100p. Q

A collection of programs written in BASIC for the Radio Shack TRS-80 or Sharp PC-1211 personal computers, but with enough information to allow modification for running on other computers using BASIC. A cassette tape with the programs recorded is available from the publisher. Includes conversion between altitude/azimuth and right ascension/declination; positional and other calculations for asteroids, comets and stars; graphic or printed ephemeris displays of Jupiter's Galilean satellites, Saturn's six brightest satellites, and the five bright planets. Appendices give further details of the programming.

363 Meeus, Jean
ASTRONOMICAL FORMULAE FOR CALCULATORS (2nd ed.)
Richmond, VA: Willmann-Bell, 1982, ISBN 0-943396-01-8 pbk. 201p; diags.

Intended for use with programmable calculators, this book is aimed at amateur astronomers with mathematical interests, and requires previous knowledge of astronomy and elementary mathematical techniques. Covers nearly 40 topics, including basics, such as interpolation, the date of Easter and other calculations of time. Also covers angular separation, planetary conjunctions, precession, nutation, apparent place of a star, the equation of time, elliptic and parabolic motion, phases of the Moon, eclipses, illuminated fraction of the Moon's disc, and of planetary discs, stellar magnitudes. The book does not provide programs, but formulae for writing them.

3 History of Astronomy

General

364 Ashbrook, Joseph, (Leif Robinson, ed.)
THE ASTRONOMICAL SCRAPBOOK: SKYWATCHERS, PIONEERS AND SEEKERS IN ASTRONOMY
Cambridge, MA: Sky, 1984, ISBN 0-933346-24-7; Cambridge: Cambridge University Press, 1984, ISBN 0-521-30045-2. 468p; illus, diags. Bibl (3), index (2).

Joseph Ashbrook (1918–80), the former editor of *Sky & Telescope* (143), contributed a series of brief articles to the magazine from 1954–80, which provided interesting insights into unusual or little known aspects of the history of astronomy. This book brings together 91 of these articles in 83 chapters, grouped in seven sections according to the topics covered.

*365 Bobrovnikoff, Nicholas T.
ASTRONOMY BEFORE THE TELESCOPE, VOL. 1: THE EARTH-MOON SYSTEM
Tucson, AZ: Pachart, 1984, ISBN 0-88126-201-3 pbk. 153p.

366 Christianson, Gale E.
THIS WILD ABYSS: THE STORY OF THE MEN WHO MADE MODERN ASTRONOMY
New York: Macmillan (The Free Press), 1979, ISBN 0-02-905380-3; 0-02-905660-8 pbk. 461 + 16p; illus. Bibl (2), index (3).

A history of astronomy up to and including the time of Newton, beginning with prehistoric man, and emphasizing the contributions of the ancient Greeks, Copernicus, Tycho Brahe, Kepler, Galileo and Newton. An attempt to produce a comprehensive work for the general reader; the author does not dwell on the personal lives of the subjects to the same extent as Koestler (371), but paints in the broader background of social and cultural developments; nor does he ignore the European Dark Ages.

For a study of Tycho Brahe, *see* John L. Dreyer, *Tycho Brahe: a picture of scientific life and work in the sixteenth century* (New York: Dover, 1963, reprint of 1st ed. 1890).

367 DeVorkin, David H.
THE HISTORY OF MODERN ASTRONOMY AND ASTROPHYSICS: A SELECTED AND ANNOTATED BIBLIOGRAPHY
New York: Garland, 1982, ISBN 0-8240-9283-X. 434p; illus. Index (3).

Contains 1,417 books and articles arranged by subject. It is intended for historians, scientists and interested readers. Includes some non-technical material, but journal articles were only included if they 'provide adequate documentation' or convey personal opinions and recollections of astronomers, but this rule is not applied consistently. Overall, however, a work of great value.

368 Gingerich, Owen (ed.)
ASTROPHYSICS AND TWENTIETH-CENTURY ASTRONOMY TO 1950: PART A (THE GENERAL HISTORY OF ASTRONOMY, VOL. 4)
Cambridge: Cambridge University Press, 1984, ISBN 0-521-24256-8. 212p; illus, diags. Bibl (x), index (2).

The purpose of this series (of which this is the only part published at the time of writing) is 'to present an account for the non-specialist of the development of astronomy from earliest times', by an international body of authors. The cut-off date is 1950. The 11 chapters in this volume cover, in part one, the birth of astrophysics in the mid-19th century, the impact of photography, and telescope building 1850–1900; and in part two: observatories and instrumentation, including separate accounts of selected astronomical institutions (generally 1850–1920), and early radio astronomy. An appendix lists 'the world's largest telescopes 1850–1950'. No special mathematical knowledge is required.

369 Hermann, Dieter B., (Kevin Krisciunas, trans. and rev.)
THE HISTORY OF ASTRONOMY FROM HERSCHEL TO HERTZSPRUNG
Cambridge: Cambridge University Press, 1984, ISBN 0-521-25733-6 (trans. of 3rd German ed.). 220p; illus, diags. Bibl (2), indexes (3).

A scholarly history covering developments in astronomy from the discovery of Neptune to the discovery of the expanding universe in 1929. Focuses on the 'principal directions' or 'main areas of activity' rather than on individual scientists or in strict chronological sequence, avoiding the artificial isolation of related aspects. The bibliography includes many technical and non-English references.

370 Hoskin, Michael
STELLAR ASTRONOMY: HISTORICAL STUDIES
Chalfont St. Giles: Science History Publications, 1982, ISBN 0-905193-04-0 pbk. 197p; illus, diags. Bibl (x), index (2).

A collection of articles, mostly reprinted from *The Journal for the History of Astronomy*, on a subject somewhat neglected in the historical study of astronomy. The introduction deals with the principles and methods of astronomy from a historical point of view. The book's three sections cover 1. the emerging science of stars, concentrating on naked-eye observations, 2. a century of speculative cosmologies, and 3. the riddle of the nebulae, ending with the 'Great Debate' (*see* 413).

371 Koestler, Arthur
THE SLEEPWALKERS: A HISTORY OF MAN'S CHANGING VISION OF THE UNIVERSE
Harmondsworth: Penguin (Pelican), 1968, ISBN 0-14-020927-7 pbk (reprinted several times; first published 1959). 623p. Bibl (3), index (3).

This fascinating, scholarly work is an account of the development of astronomy and cosmology up to the time of Newton, and also the story of the intellectual changes which accompanied the 'scientific revolution', conveyed by focusing on the lives and work of Copernicus (*see also* 430), Kepler, and Galileo (*see also* 434 and 435).

Two biographies of Kepler are, Max Caspar, *Kepler* (London and New York: Abelard-Schuman, 1959 (trans. of 1st German ed. 1948, 401p); and Angus Armitage, *John Kepler* (London: Faber, 1966; New York: Roy, 1966, 194p).

372 Lang, Kenneth R. and Owen Gingerich (ed.)
A SOURCE BOOK IN ASTRONOMY AND ASTROPHYSICS, 1900–1975
Cambridge, MA: Harvard University Press, 1979, ISBN 0-674-82200-5. 922p; illus, diags. Indexes (3).

This volume aims to show the development of astronomy in the 20th century to 1975, through the original publications of important astronomers. The 132 selections are mostly from scholarly and professional journals, but the level varies from the popular to the highly technical. The articles are arranged chronologically within eight chapters, which follow the traditional subject arrangement of general textbooks.

373 Loomis, Elias, (Bernard I. Cohen, ed.)
THE RECENT PROGRESS OF ASTRONOMY: ESPECIALLY IN THE UNITED STATES
Salem, NH: Arno, 1980, ISBN 0-405-12554-2 (facsimile reprint of 1st ed. 1851). 257p; diags. Bibl (x).

'A popular account of some of the most important astronomical discoveries which took place in the 1840–50 decade'. The most noteworthy event covered is the discovery of Neptune. (3rd ed. 1856 only seen).

374 Moore, Patrick
THE DEVELOPMENT OF ASTRONOMICAL THOUGHT (2nd ed.)
Hornchurch: Ian Henry, 1980, ISBN 0-86025-841-6. 108p; illus, diags.

A brief account of mankind's changing outlook on the universe, adapted by the discoveries of astronomy.

375 Moore, Patrick
PATRICK MOORE'S HISTORY OF ASTRONOMY (6th ed.)
London: Macdonald, 1983, ISBN 0-356-08607-0. 327p; illus & diags (mostly col.). Index (2). Q

A profusely illustrated history, from prehistoric times to the present, for the general reader. This edition has been substantially revised and enlarged; it has 75 pages more than the fifth edition. (The first three editions were titled *Astronomy*; this was changed for the fourth and fifth editions to *The story of astronomy*).

376 Murdin, Lesley
UNDER NEWTON'S SHADOW: ASTRONOMICAL PRACTICES IN THE SEVENTEENTH CENTURY
Bristol and Accord, MA: Adam Hilger, 1985, ISBN 0-85274-456-0. 152p; illus, diags. Bibl (2), index (1).

A survey of the characters and personal lives of the astronomers of the 17th and early 18th century, who were stimulated by Isaac Newton's theories of gravity and motion. In addition to prominent figures such as Hooke, Halley and Flamsteed, the book also covers less well known figures such as Stephen Gray and Thomas Brattle.

377 Tauber, Gerald E.
MAN'S VIEW OF THE UNIVERSE: A PICTORIAL HISTORY
New York: Greenwich Press, 1982, ISBN 0-517-38769-9 (reprint of 1st ed. 1979). 352p; illus, diags. Bibl (1), index (2). Q

This book provides a basic introduction to astronomy, but through a completely historical approach. The text moves from prehistory to the present, ending with a chapter on space exploration.

378 Wilson, Colin
STARSEEKERS
London: Hodder & Stoughton, 1980, ISBN 0-340-25859-4; New York: Doubleday, 1980, ISBN 0-385-17253-2; London: Granada (Panther), 1982, ISBN 0-586-05575-4 pbk. 271p; illus (some col.); 336p (UK pbk). Bibl (2), index (2).

A fairly brief history of man's discovery of the universe from prehistoric times to the present. A wide-ranging account which touches on religion, philosophy, mathematics and other sciences, by an author who takes unorthodox theories more seriously than most other writers of astronomical histories.

379 SOME EARLIER GENERAL HISTORIES OF ASTRONOMY

Berry, Arthur
A short history of astronomy from earliest times through the 19th century (New York: Dover, 1961, ISBN 0-486-20210-0 pbk (reprint of 1st ed. 1898), 440p).

Dreyer, J.L.E.
A history of astronomy from Thales to Kepler (New York: Dover, 1953, ISBN 0-486-60079-3 pbk (reprint of *History of planetary systems from Thales to Kepler*, 1906), 438p).

Ley, Willy
Watchers of the skies: an informal history of astronomy from Babylon to the space age (New York: Viking, 1963; London: Sidgwick & Jackson, 1964, 528p).

Pannekoek, A.
A history of astronomy (London: Allen & Unwin, 1961; Totowa, NJ: Rowman & Littlefield, 1961, ISBN 0-87471-365-X, 521p). (The standard work on the subject).

Ronan, Colin A.
Astronomy ('Illustrated Sources in History') (Newton Abbot: David & Charles, 1973, ISBN 0-7153-5593-7; New York: Barnes & Noble, 1973, ISBN 0-06-495970-8, 112p).

Ronan, Colin A.
Discovering the universe: a history of astronomy (New York: Basic Books, 1971, ISBN 0-465-01670-7; London: Heinemann, 1972, ISBN 0-435-68680-1, 248p).

Thiel, Rudolf
And there was light: the discovery of the universe (New York: Knopf, 1957; London: André Deutsch, 1958 (trans. of 1st German ed. 1956), 396p).

Archaeoastronomy

380 Aveni, Anthony F. (ed.)
ARCHAEOASTRONOMY IN THE NEW WORLD
Cambridge: Cambridge University Press, 1982, ISBN 0-521-24731-4. 219p; illus, diags. Bibl (1).

The proceedings of a conference on American primitive astronomy held at Oxford University in September 1981. Somewhat technical, but should interest the historian or serious general reader. The book was reproduced from typescript and, unfortunately, lacks an index.

381 Aveni, Anthony F.
SKYWATCHERS OF ANCIENT MEXICO
Austin, TX: University of Texas Press, 1980, ISBN 0-292-77557-1. 355 + 1p; illus (1 col.), diags. Bibl (3), index (3).

This scholarly work may appeal to the historian or the general reader with a serious interest in the subject.

382 Brecher, Kenneth and Michael Feirtag (ed.)
ASTRONOMY OF THE ANCIENTS
Cambridge, MA: MIT Press, 1979, ISBN 0-262-02137-4. 206p; illus, diags. Bibl (x), index (2).

A collection of eight articles for the general reader, originally published in a special issue of *Technology Review* (Dec. 1977). The editors hope to show how ancient descriptions, in whatever form, can aid modern astronomers' understanding of events such as the birth of the Crab Nebula, or phenomena such as sunspots.

383 Brennan, Martin
THE STARS AND THE STONES: ANCIENT ART AND ASTRONOMY IN IRELAND
London: Thames & Hudson, 1983, ISBN 0-500-01295-4. 208p; illus, diags. Bibl (2), index (1).

Intended for the general reader, this book presents the results of an investigation by the author and his colleagues of megalithic sites in Ireland. Most of the sites were found to be linked with the position of the Sun at critical times of the year. There was also a link between megalithic art and astronomical events. The book includes about 300 photographs and drawings, some in two colours.

384 Burl, Aubrey
PREHISTORIC ASTRONOMY AND RITUAL
Aylesbury: Shire, 1983, ISBN 0-85263-621-0 pbk. 56p; illus, diags. Glos (1), bibl (1), index (1).

A brief look at some megalithic sites in the British Isles. The author is not so much concerned with the mechanics of prehistoric observatories, but with the ancient people and whether they aligned their monuments in accordance with lunar and solar events, and how the archaeoastronomer might attempt to discover this.

385 Cornell, James
THE FIRST STARGAZERS: AN INTRODUCTION TO THE ORIGINS OF ASTRONOMY
New York: Scribner's, 1981, ISBN 0-684-16799-9; London: Athlone Press, 1981, ISBN 0-485-30004-4. 262p; illus. Bibl (2), index (2).

Intended for the general reader, this book aims to summarize the voluminous material about the growing science of archaeoastronomy in a single comprehensive overview. Describes the development of this subject, which attempts to discover the astronomical practices of ancient peoples through the archaeological study of their artefacts, and discusses findings related to ancient peoples all over the world.

386 Hadingham, Evan
EARLY MAN AND THE COSMOS
London: Heinemann, 1983, ISBN 0-434-31108-1; New York: Walker, 1984, ISBN 0-8027-0745-9; Norman, OK: University of Oklahoma Press, 1985, ISBN 0-8061-1919-5 pbk. 277p; illus, diags. Bibl (3), glos (2), index (2).

Written by an archaeologist, this book aims chiefly 'to demonstrate the remarkable connections that existed between the observations of the skywatchers and the sacred and everyday aspects of their lives', rather than taking astronomical lore in isolation. Avoids technical terms and discussion of the modern understanding of astronomy. Concentrates on north-western Europe, North America and the Maya, ancient and modern.

387 Heggie, Douglas C. (ed.)
ARCHAEOASTRONOMY IN THE OLD WORLD
Cambridge: Cambridge University Press, 1982, ISBN 0-521-24734-9. 280p; illus, diags. Bibl (x), index (2).

A collection of 16 scholarly papers presented at a symposium held in Oxford, September 1981, which should be accessible to any seriously interested reader. The authors deal with the question of whether astronomy was practised in prehistoric north-western Europe, and, if so, why? The contributors include astronomers, archaeologists, mathematicians and computer scientists.

388 Hoyle, Fred
ON STONEHENGE
London: Heinemann, 1977, ISBN 0-435-32958-8; San Francisco: Freeman, 1977, ISBN 0-7167-0364-5; 0-7167-0363-7 pbk. 128p; illus, diags. Bibl (x), index (1).

A review of the evidence for the astronomical interpretation of Stonehenge. Includes a mathematical appendix.

The pioneer in this field was Gerald S. Hawkins; his books are *Stonehenge decoded* (London: Souvenir Press, 1966), and *Beyond Stonehenge* (London: Hutchinson, 1973, ISBN 0-09-117900-9).

389 Krupp, E.C.
ECHOES OF THE ANCIENT SKIES: THE ASTRONOMY OF LOST CIVILIZATIONS
New York: Harper & Row, 1983, ISBN 0-06-015101-3; New York: New American Library (Meridian), 1984, ISBN 0-452-00679-1 pbk. 386p; illus, diags. Bibl (3), index (2).

This wide-ranging survey of archaeoastronomy for the general reader is arranged by theme rather than by the different ancient peoples covered. The 13 chapters discuss such topics as ancient and prehistoric observatories, sky gods, mythology, calendars, temples, symbols and cosmology. Includes a comprehensive bibliography.

390 Krupp, E.C. (ed.)
IN SEARCH OF ANCIENT ASTRONOMIES
Garden City, NY: Doubleday, 1978, ISBN 0-385-11639-X: New York: McGraw-Hill, 1979, ISBN 0-07-035556-8 pbk; London: Chatto & Windus, 1979, ISBN 0-7011-2314-1; Harmondsworth: Penguin, 1984, ISBN 0-14-005481-2 pbk. 277p; illus, diags. Bibl (3), index (2).

A collection of articles intended for the general reader. The editor commments that the study of the astronomies of ancient and prehistoric times brings new light to our knowledge of the achievements of ancient peoples, and provides a useful corrective to extravagant claims about ancient astronauts. Covers Stonehenge, and the astronomy of selected peoples all over the world, such as American Indians and the Egyptians. Does not cover ancient cultures such as Mesopotamia, China and Greece, where our knowledge of their astronomy is mainly from written records. The final chapter discusses unorthodox and pseudoscientific theories. (The 1984 edition has a revised bibliography).

391 Lancaster-Brown, Peter
MEGALITHS AND MASTERMINDS
London: Robert Hale, 1979, ISBN 0-7091-0915-9; New York: Scribner's, 1979, ISBN 0-684-16891-X. 246 + 16p; illus, diags. Bibl (1), index (2).

A popular survey of our knowledge of megaliths, including the pyramids, Stonehenge, and others. Discusses theories regarding their possible astronomical significance, and looks at the dispute between archaeologists who argued against an astronomical connection, and the proponents of astronomical theories.

392 Michell, John
A LITTLE HISTORY OF ASTRO-ARCHAEOLOGY: STAGES IN THE TRANSFORMATION OF A HERESY (UK)/ SECRETS OF THE STONES: THE STORY OF ASTRO-ARCHAEOLOGY (US)
London: Thames & Hudson, 1977, ISBN 0-500-01183-4; New York: Penguin, 1977, ISBN 0-14-004491-4 pbk. 96p; illus, diags. Bibl (1), index (1).

A highly illustrated essay, 'designed to illustrate the stages by which a new idea, in this case the theory behind astro-archaeology, promotes itself in status from lunacy to heresy to interesting notion and finally to the gates of orthodoxy'. The notion being that prehistoric peoples had astronomical knowledge far beyond what they had been credited with by archaeologists, a view contrary to the generally 'currently held belief that civilization is a recent phenomenon'.

*393 Piini, Ernest W.
AMERICA'S STONEHENGE
Redwood City, CA: Sarsen Press, 1980, pbk. 29p.

Constructed at Maryhill in the state of Washington, the American replica of Stonehenge was an eccentric project of millionaire Sam Hill. It was completed in 1929. The author gives background information, and reports his success in making alignment measurements for the various risings and settings of the Sun and Moon. (From *S&T* Dec. 1981, p 595).

394 Urton, Gary
AT THE CROSSROADS OF THE EARTH AND THE SKY
Austin, TX: University of Texas Press, 1981, ISBN 0-292-70349-X. 248p; diags. Bibl (3), index (3).

A scholarly study of Andean astronomy, in particular that of the Quechua-speaking Indians. This is a study in ethnoastronomy rather than archaeoastronomy; the author uses such methods as textual scholarship, ethnology, and the interpretation of ancient iconography to reconstruct the astronomical techniques of his subjects.

395 Williamson, Ray A.
LIVING THE SKY: THE COSMOS OF THE AMERICAN INDIAN
Boston: Houghton Mifflin, 1984, ISBN 0-395-35414-5. 366 + 24p; illus (some col.), diags. Bibl (3), index (3).

A scholarly survey of native American cosmology which will suit any reader interested in this subject.

396 Wood, John Edwin
SUN, MOON AND STANDING STONES (2nd ed.)
Oxford: Oxford University Press, 1980, ISBN 0-19-285089-X pbk. 217p; illus, diags. Bibl (2), index (2).

A survey of archaeoastronomical findings in north-western Europe for non-specialists, which assumes no previous knowledge of astronomy.

For a discussion of the ideas behind the concept of ancient astronomy, *see* Douglas C. Heggie's *Megalithic science: ancient mathematics and science in north-west Europe* (London: Thames & Hudson, 1981, ISBN 0-500-27211-5 pbk).

Specific Topics

The telescope

397 Asimov, Isaac
EYES OF THE UNIVERSE: A HISTORY OF THE TELESCOPE
Boston: Houghton Mifflin, 1975, ISBN 0-395-19427-X; London:

André Deutsch, 1976, ISBN 0-233-96760-5; London: Quartet, 1978, ISBN 0-7043-3197-7 pbk. 274 + 16p; illus. Index (2).

A readable account of the development of the telescope as an astronomical tool, and the discoveries made with it. This wide-ranging discussion of astronomy provides a basic introduction to the field, as well as taking the story of the telescope beyond its use in the visible part of the spectrum.

398 King, Henry C.
THE HISTORY OF THE TELESCOPE
New York: Dover, 1979, ISBN 0-486-23893-8 pbk (reprint of 1st ed. 1955). 486p; illus, diags. Bibl (x), index (3).

This scholarly work, with hundreds of references to original sources, was the first attempt to present a comprehensive history of the telescope. The author adds interest by providing details of the craftsmen and instrument makers who advanced the art of telescope making, and by noting associated developments in astronomy.

399 Learner, Richard
ASTRONOMY THROUGH THE TELESCOPE
New York: Van Nostrand, 1981, ISBN 0-442-25839-9; London: Evans, 1982, ISBN 0-237-45644-3. 224p; illus (some col.), diags. Glos (2), index (3). Q

A beautifully illustrated history of the telescope for anyone interested in the subject. Tells the story from Galileo to radio telescopes, with notes on current and future projects. Includes a chapter on amateur astronomy.

400 van Helden, Albert
THE INVENTION OF THE TELESCOPE
Philadelphia: American Philosophical Society, 1977, pbk. 67p; illus, diags. Bibl (x), index (2).

First published as *Transactions of the American Philosophical Society* (vol. 67 (4), 1977), this is a brief scholarly essay, with extensive footnotes, followed by a collection of pertinent documents in their original language and in English translation.

Places

401 Brück, Hermann A.
THE STORY OF ASTRONOMY IN EDINBURGH: FROM ITS BEGINNINGS UNTIL 1975
Edinburgh: Edinburgh University Press, 1983, ISBN 0-85224-480-0. 151p; illus. Bibl (2).

The history of astronomy at the University and the Royal Observatory, with biographical sketches of Edinburgh astronomers. The book's three parts cover the early history and the era of Charles Piazzi Smyth; the new observatory at Blackford Hill, 1888–1957; and 1957–75, the 'time of new technology and change'. Unfortunately there is no index.

*402 Kloeppel, James E.
REALM OF THE LONG EYES: A BRIEF HISTORY OF KITT PEAK NATIONAL OBSERVATORY
San Diego, CA: Univelt, 1982, ISBN 0-912183-01-2 pbk. 136p; illus.

This account focuses on the establishment of the facility and the design and building of its telescopes. The observatory's former space division is discussed at length, but there is a lack of attention to the work and achievements of the Kitt Peak telescopes. (From *S&T* July 1984, p 35).

403 McCrea, W.H.
THE ROYAL GREENWICH OBSERVATORY
London: HMSO, 1975, ISBN 0-11-880616-5 pbk. 80 + 12p; illus (some col.). Bibl (1).

The story of the RGO from its founding in 1675, through its move to Herstmonceux in the 1950s, to its more recent work (up to 1975). Discusses the roles played by the various Astronomers Royal and their staff, the instruments used, and the work carried out there. Ends with a chronology which includes background astronomical events.

404 Moore, Patrick
THE ASTRONOMY OF BIRR CASTLE (2nd ed.)
Birr, Ireland: Tribune, 1981, pbk. 81 + 9p; illus, diags. Bibl (1).

The history of astronomy at Birr Castle in Ireland, beginning in the 1840s with the 3rd Earl of Rosse, who built the great 72-inch reflector, and whose work was carried on by his successors. (The second edition is only slightly revised).

405 Moore, Patrick and Pete Collins
THE ASTRONOMY OF SOUTHERN AFRICA
Cape Town: H. Timmins, 1977, ISBN 0-86978-143-X; London: Robert Hale, 1977, ISBN 0-7091-6176-X. 160p; illus (some col.), diags. Index (2). Q

The history of astronomy in Southern African, beginning in 1685 with the arrival of a French Jesuit priest, Father Guy Tachard, who established the first observatory at the Cape of Good Hope (*see also*

407). Includes some coverage of amateur astronomy.

406 Rochester, G.D.
THE HISTORY OF ASTRONOMY IN THE UNIVERSITY OF DURHAM FROM 1835–1939
Durham: University of Durham, 1980, pbk. 10p.

This pamphlet is a reprint of a paper which appeared in the *QJRAS* (vol. 21, 1980, p 369–78).

407 Warner, Brian
ASTRONOMERS AT THE ROYAL OBSERVATORY CAPE OF GOOD HOPE: A HISTORY WITH EMPHASIS ON THE NINETEENTH CENTURY
Cape Town: A.A. Balkema, 1979, ISBN 0-86961-109-7. 132p; illus, diags. Bibl (1), index (1).

A history based 'solely on original sources'; intended to reach a wide readership, so the text is not encumbered with references. The story, as the title suggests, focuses on the personalities involved, those at the Cape and notable visitors. The illustrations include samples of the work of artists who have been attracted to the Cape.

Discoveries

408 Berendzen, Richard, Richard Hart and Daniel Seeley
MAN DISCOVERS THE GALAXIES
New York: Columbia University Press (Morningside Books), 1984, ISBN 0-231-05826-8; 0-231-05827-6 pbk (reprint of 1st ed. 1976). 228p; illus, diags. Bibl (1), index (1).

In this book, which is intended for 'any reader with an interest in galactic astronomy', the authors hope to throw light on the ways of science and scientists through the study of one aspect of science in detail. This aspect, the history of the discovery of galaxies as a class of celestial object, is surveyed using original sources, many previously unpublished. Includes questions. (This reprint contains a new preface).

For the Milky Way, a readable, rather personal account is Charles A. Whitney's *The discovery of our galaxy* (New York: Knopf, 1971, ISBN 0-394-46068-5; London: Angus & Robertson, 1972, ISBN 0-207-95452-6, 317p).

409 Grosser, Morton
THE DISCOVERY OF NEPTUNE
New York: Dover, 1979, ISBN 0-486-23726-5 pbk (reprint of 1st ed. 1962). 172p. Glos (2), bibl (3), index (2).

Following a survey of planetary astronomy to 1781, two chapters discuss the discovery of Uranus and the problems over the discrepancies between its observed and predicted motions. The next two chapters deal with Jean Joseph Leverrier and John Couch Adams, who used the discrepancies to predict the presence of a planet beyond Uranus. The final chapters cover the dispute over priority in the predictions, which arose in astronomical circles, and the actual discovery in 1846. (This reprint contains a new preface by the author).

410 Hoyt, William Graves
PLANETS X AND PLUTO
Tucson, AZ: University of Arizona Press, 1980, ISBN 0-8165-0664-7 pbk. 302p; illus. Bibl (3), index (3).

This scholarly work presents a good account for anyone interested in the subject of planetary searching. Begins with the discovery of Uranus and the larger asteroids, which fitted nicely into the scale of planetary distances of Bode's Law, thus sparking three successive searches for planets X. Neptune and Pluto were subsequently found, but the second planet X, Vulcan, supposedly inside the orbit of Mercury, was not.

411 Hubble, Edwin
THE REALM OF THE NEBULAE
New Haven and London: Yale University Press, 1982, ISBN 0-300-02499-1; 0-300-02500-9 pbk (reprint of 1st ed. 1936). 207p; illus, diags. Index (2).

Republication of a classic work, based on lectures given in 1935, in which Hubble described his discovery of the expanding nature of the universe, based on the evidence of Cepheid variable stars used as distance indicators to show that some supposed nebulae were receding too rapidly to be part of our galaxy, and were in fact galaxies themselves.

412 Jones, Kenneth Glyn
THE SEARCH FOR THE NEBULAE
Chalfont St. Giles: Alpha Academic, 1975, 84p. Bibl (1).

Revised and corrected from its first appearance as a series of articles in the *JBAA*, this book 'attempts to provide a brief history of the early discoveries of nebulae and star clusters, and, in particular to present a comprehensive analysis of the several catalogues dealing with these objects up to the time of William Herschel'. Arranged in short chapters each dealing with one astronomer. Includes extensive quotations from original sources.

413 Smith, Robert W.
THE EXPANDING UNIVERSE: ASTRONOMY'S GREAT DEBATE 1900–1931
Cambridge: Cambridge University Press, 1982, ISBN 0-521-23212-0. 220p; illus, diags. Bibl (3), index (2).

This scholarly work, intended for professional and amateur astronomers and science historians, incorporates the results of much original research. Covers the period which saw three radical changes in the generally accepted view of the universe: 1. an approximately tenfold increase in the size of the galaxy, 2. the acceptance of the existence of external galaxies, and 3. the realization that these galaxies disclose the expanding nature of the universe. The book is essentially concerned with intellectual factors, but the author points out that factors such as sociology may well be necessary to a full understanding of the intellectual shifts well illustrated by the great debate between Harlow Shapley and H.P. Curtis on the scale of the universe.

414 Tombaugh, Clyde W. and Patrick Moore
OUT OF THE DARKNESS: THE PLANET PLUTO
Guildford: Lutterworth Press, 1980, ISBN 0-7188-2500-4; Harrisburg, PA: Stackpole, 1980, ISBN 0-8117-1163-3. 221p; illus, diags. Glos (2), index (2).

The story of the discovery of Pluto by the man who was responsible for the painstaking work involved. The five (of 16) chapters by Moore deal with the historical background, and with what little is known about the planet.

415 van Helden, Albert
MEASURING THE UNIVERSE: COSMIC DIMENSIONS FROM ARISTARCHUS TO HALLEY
Chicago and London: University of Chicago Press, 1985, ISBN 0-226-84881-7. 203p; diags. Bibl (3), index (2).

A scholarly history, with extensive notes, of the development of our understanding of cosmic dimensions up to the 17th century, when astronomers began to grasp the true scale of the universe.

New astronomies

416 Hirsch, Richard F.
GLIMPSING AN INVISIBLE UNIVERSE: THE EMERGENCE OF X-RAY ASTRONOMY
Cambridge: Cambridge University Press, 1983, ISBN 0-521-25121-4;

1985, ISBN 0-521-31232-9 pbk. 186p; illus, diags. Bibl (3), index (2).

A brief, but extensively researched history covering the field's initial phase of development from the 1950s to the early 1970s, when the basic data and theoretical principles were established. There is a US bias as the author shows the role of technical advances and public policy changes in the development of x-ray astronomy.

*417 Kellerman, K. and B. Sheets (ed.)
SERENDIPITOUS DISCOVERIES IN RADIO ASTRONOMY
Green Bank, WV: National Radio Astronomy Observatory, 1984, pbk. 321p.

These proceedings of a workshop on serendipity make up a history of radio astronomy from the early work of Karl Jansky up to the discovery of pulsars. All the comments and arguments which accompanied the formal presentations have been left in to present a realistic picture of how science is *really* done. (From *S&T* March 1985, p 228).

418 Lovell, Bernard
THE JODRELL BANK TELESCOPES
Oxford: Oxford University Press, 1985, ISBN 0-19-858178-5. 292p; illus, diags. Bibl (3), index (3).

This book is concerned not with the results of astronomical research, but with the background to the development of ideas, and the financial and political background to the building of large, expensive equipment. Covers the period 1960–82, taking in not only those projects which came to fruition, but also those instruments which might have existed in more favourable political and economic circumstances. The book complements Lovell's *The story of Jodrell Bank* (London: Oxford University Press, 1968) and *Out of the zenith: Jodrell Bank 1957–1970* (London: Oxford University Press, 1973, ISBN 0-19-217624-2), the second of which was concerned mainly with astronomical researches rather than the development of instrumentation. Appendices include a list of Jodrell Bank steerable telescopes, proposed and actual. (*See also* 444).

419 Sullivan, W.T., III (ed.)
THE EARLY YEARS OF RADIO ASTRONOMY: REFLECTIONS FIFTY YEARS AFTER JANSKY'S DISCOVERY
Cambridge: Cambridge University Press, 1984, ISBN 0-521-25485-X. 421p; illus, diags. Bibl (x), indexes (2).

The 21 chapters in this book bring together contributions from many of the field's pioneers in papers based on half-day sessions at three major conferences in 1980, 1981 and 1982. Technical in parts, the book generally deals with the development of radio astronomy before 1960, but it is not a formal history, more an attempt to convey the spirit of the times, to complement historical study and 'provide starting points for delving further into the history'.

A historical account by a pioneer in the field is J.S. Hey's *The evolution of radio astronomy* (London: Paul Elek, 1973, ISBN 0-236-15453-2; Nantucket, MA: Neale Watson, 1973, ISBN 0-88202-027-7). A book which brings a sociological perspective to the subject is, David Edge and Michael J. Mulkay, *Astronomy transformed: the emergence of radio astronomy in Britain* (New York: Wiley, 1976, ISBN 0-471-23273-4, 482p). (*See also* 438).

420 Tucker, Wallace and Riccardo Giacconi
THE X-RAY UNIVERSE
Cambridge, MA and London: Harvard University Press, 1985, ISBN 0-674-96285-0. 201 + 4p; illus (some col.), diags. Bibl (2), index (2).

A history of x-ray astronomy for the general reader, by one of the field's pioneers (Giacconi) and a theorist who has worked closely with him. The authors include many background details of the people involved, and provide good coverage of associated astronomical and cosmological advances.

Others

421 Allen, Richard Hinckley
STAR NAMES: THEIR LORE AND MEANING
New York: Dover, 1963, ISBN 0-486-21079-0 pbk. 563p. Bibl (2), indexes (3).

Drawing extensively on early records, the author provides an exhaustive survey of star names. The book does not take the form of a dictionary, but is a descriptive text with the names of stars shown in bold type. The main part of the book (464 pages) covers the constellations in alphabetical order. (This is an unabridged, corrected reprint of *Star-names and their meanings*, 1899). Not listed by Seal (24).

422 Drake, Stillman
TELESCOPES, TIDES AND TACTICS: A GALILEAN DIALOGUE ABOUT THE *STARRY MESSENGER* AND SYSTEMS OF THE WORLD
Chicago: University of Chicago Press, 1983, ISBN 0-226-16231-1. 236p; diags. Index (2).

Presents Galileo's *Starry messenger* of 1610 in the context of an imagined (somewhat dry) dialogue among three of his friends late in 1613. A companion to the author's *Cause, experiment and science* (University of Chicago Press, 1984, ISBN 0-226-16230-3) which deals with Galileo's first book on physics in a similar way. Supplements (435) with new material drawn from letters and notebooks.

423 Gallant, Roy A.
THE CONSTELLATIONS: HOW THEY CAME TO BE
New York: Four Winds Press, 1979, ISBN 0-590-07552-7. 203p; illus, diags. Glos (2), index (2).

Aimed primarily at younger readers, this book provides a good basic introduction to the lore of the constellations for anyone. Includes finder charts.

424 Jaki, Stanley L.
PLANETS AND PLANETARIANS: A HISTORY OF THEORIES OF THE ORIGIN OF PLANETARY SYSTEMS
Edinburgh: Scottish Academy Press, 1978, ISBN 0-7073-0135-5. 266p; diags. Index (2).

A rather dry scholarly work with many references to original sources.

425 King, Henry C. and John R. Millburn
GEARED TO THE STARS: THE EVOLUTION OF PLANETARIUMS, ORRERIES AND ASTRONOMICAL CLOCKS
Bristol: Adam Hilger, 1978, ISBN 0-85274-399-8; Toronto: University of Toronto Press, 1978, ISBN 0-8020-2312-6. 442p; illus, diags. Glos (2), bibl (3), index (3). Q

A large scholarly tome, likely to interest only those with a serious interest in this subject.

426 Morphet, Clive
GALILEO AND COPERNICAN ASTRONOMY: A SCIENTIFIC WORLD VIEW DEFINED
London: Butterworths, 1977, ISBN 0-408-71303-8 pbk. 80p; diags. Bibl (2).

This book is designed for 'a course intended to explore the nature of scientific ideas, and the way in which they interact with society by focusing on the struggle that Galileo Galilei had with the church over the Copernican theory of the universe'. Includes extracts from secondary sources, summaries, questions and teacher's notes, but no index.

427 Snyder, George Sergeant
MAPS OF THE HEAVENS
New York: Abbeville Press, 1984, ISBN 0-89659-456-4; London: André Deutsch, 1984, ISBN 0-233-97721-X. 144p; illus (mostly col.). Bibl (1). Q

A brief history of the development of celestial cartography, illustrated with many large, full-page examples. The maps presented and described were selected to represent different artistic traditions, and different media and techniques of celestial cartography.

428 Warner, Deborah J.
THE SKY EXPLORED: CELESTIAL CARTOGRAPHY 1500–1800
New York: Alan R. Liss, 1979, ISBN 0-8451-1700-9. 293p; illus. Bibl (1), index (3). Q

This book 'aims to make star maps intelligible to a wide audience, many of whom may not be astronomers or cartographers'. The introduction gives background information and an explanation of technical terms, but the bulk of the book is a catalogue of flat maps, alphabetically arranged by cartographer, with a reproduction of the whole, or part of the map, giving basic details and notes.

Biography

429 Abbott, David (ed.)
THE BIOGRAPHICAL DICTIONARY OF SCIENTISTS: ASTRONOMERS
London: Frederick Muller (Blond Educational), 1984, ISBN 0-584-70002-4; New York: Peter Bedrick, 1984, ISBN 0-911745-80-7. 204p; illus, diags. Glos (3), index (3).

Part of a series covering the traditional major scientific disciplines, intended primarily for school and college students, but also for the general reader. The short historical introduction is followed by about 200 alphabetically arranged biographical sketches, mostly one page or less in length. Coverage extends from the earliest astronomers, such as Thales and Anaximander, to those still working, such as Carl Sagan and Stephen Hawking.

430 Armitage, Angus
THE WORLD OF COPERNICUS
Wakefield: EP Publishing, 1972, ISBN 0-85409-737-6; Woodstock, NY: Beekman, 1972, ISBN 0-8464-0979-8 (reprint of 1st ed. 1947). 165 + 6p; illus, diags. Bibl (1), index (1).

Describes the life and work of the man, and the triumph of the Copernican view of the Solar System. (Originally published as *Sun, stand thou still*, the present title was given to the 1951 printing). Not listed by Seal (24).

For a different perspective compare this work with John Banville's biographical novel, *Doctor Copernicus* (1976). A further novel about a key figure in the history of astronomy is Banville's *Kepler* (1981). *See also* Fred Hoyle, *Nicolaus Copernicus: an essay on his life and work* (London: Heinemann, 1973, ISBN 0-435-54425-X, 84p).

431 Chandrasekhar, S.
EDDINGTON: THE MOST DISTINGUISHED ASTROPHYSICIST OF HIS TIME
Cambridge: Cambridge University Press, 1983, ISBN 0-521-25746-8. 64p; illus, diags. Bibl (2), index (1).

An assessment of Eddington's contribution to astronomy and astrophysics based on two lectures given in Cambridge to mark the centenary of his birth. Largely non-technical, the book is aimed primarily at astronomers, mathematicians and historians of modern astronomy.

432 Chapman, Allan
THREE NORTH COUNTRY ASTRONOMERS
Manchester: Neil Richardson, 1982, ISBN 0-907511-09-0 pbk. 41p; illus, diags. Bibl (1).

The story of William Gascoigne, Jeremiah Horrocks and William Crabtree, and their fundamental discoveries in astronomy and the design of scientific instruments. Regrettably, their group broke up in the early 1640s with the premature deaths of its members.

433 Christianson, Gale E.
IN THE PRESENCE OF THE CREATOR: ISAAC NEWTON AND HIS TIMES
New York: Macmillan (The Free Press), 1984, ISBN 0-02-905190-8. 623p; illus, diags. Bibl (3), index (3).

This extensively researched biography invokes Newton's own voice whenever possible and does not gloss over the more difficult parts of his life. The author's purpose is 'to bring Newton to life rather than to embalm him'. (*See also* 447).

434 Drake, Stillman
GALILEO
Oxford: Oxford University Press, 1980, ISBN 0-19-287527-2; 0-19-287526-4 pbk. 100p. Bibl (1), index (2).

A brief introduction to Galileo's life and work, focusing on his condemnation by the Roman Inquisition in 1633.

435 Drake, Stillman
GALILEO AT WORK: HIS SCIENTIFIC BIOGRAPHY
Chicago: University of Chicago Press, 1978, ISBN 0-226-16226-5. 536p; illus, diags. Bibl (1), index (3).

The author intends to show Galileo as a working scientist, as well as a profound scholar, the aspect of his life to which most other biographies devote themselves. Follows a strict chronological track in an attempt to 'find out what things actually shaped his thought, as distinguished from those that ought to have shaped it as determined by probable reasoning'. Includes a 35-page appendix of brief biographical sketches of people mentioned in the text.

See also Colin A. Ronan, *Galileo* (London: Weidenfeld & Nicolson, 1974, ISBN 0-297-76801-8; New York: Putnam, 1974, ISBN 0-399-11364-9, 264p).

436 Fernie, Donald
THE WHISPER AND THE VISION
Toronto: Clarke, Irwin, 1976, ISBN 0-7720-1090-0. 189p; illus. Bibl (1).

An informal excursion into the history of astronomy, which recounts the experiences and adventures of travelling astronomers in the 18th and 19th centuries. The first chapter, "Transits and tribulations", describes the efforts of various astronomers to get to the right place at the right time. Three further chapters deal with Maclear and Herschel in Southern Africa, David Gill in search of the solar parallax, and Solon Bailey in South America. Not listed by Seal (24).

437 Hoyt, William Graves
LOWELL AND MARS
Tucson, AZ: University of Arizona Press, 1976, ISBN 0-8165-0514-4 pbk. 376p; illus, diags. Bibl (2), index (2).

A scholarly, but readable account of the life of Percival Lowell, concentrating on his work and ideas, primarily those concerned with the planet Mars. (*See also* 688). Not listed by Seal (24).

438 Kraus, John
BIG EAR
Powell, OH: Cygnus-Quasar, 1976. 228p; illus, diags. Bibl (2), index (2).

A highly individualistic account by a pioneer of radio astronomy of his involvement in the field. He recounts the story of the construction of his own equipment in Ohio, culminating in the building of 'Big Ear' and its use in detecting radio signals from space. Includes a five-page listing of papers by Kraus, and two pages of general references. Not listed by Seal (24).

*439 Marshall, Oscar Seth, (Eva M. Douglas, ed.)
JOURNEYMAN MACHINIST EN ROUTE TO THE STARS
Taunton, MA: W.S. Sullwold, 1979, ISBN 0-88492-025-9. 156p.

The autobiography of a Vermont amateur (1870–1953), who tells the story of his long association with Russell W. Porter and the Springfield Telescope Makers, and how this led to a share in the construction of the 200-inch Hale telescope. (From *S&T* Jan. 1980, p 62).

For a biography of Porter, *see* Willard C. Berton, *Russell W. Porter: Arctic explorer, artist and telescope maker* (Freeport, ME: Cumberland Press, 1976, ISBN 0-87027-168-7, 274p).

440 Moore, Patrick
WILLIAM HERSCHEL: ASTRONOMER AND MUSICIAN OF 19 NEW KING STREET, BATH
Sidcup: P.M.E. Erwood, 1981, ISBN 0-907322-06-9 pbk. 28p; illus, diags. Bibl (1).

A booklet written by the president of the William Herschel Society to commemorate the anniversary of the discovery of Uranus in 1781. Attempts to give a brief overall impression of the man himself, with particular reference to his associations with the city of Bath, where Herschel was living when he made the discovery. (*See also* 701).

A fuller biography of Herschel is Angus Armitage's *William Herschel* (London: Thomas Nelson, 1962; New York: Fernhill, 1962, 158p). For his son, *see* Günther Buttman, *The shadow of the telescope: a biography of John Herschel* (New York: Scribner's, 1970; Guildford: Lutterworth Press, 1974, ISBN 0-7188-2087-8, 219p).

441 Osterbrock, Donald E.
JAMES E. KEELER, PIONEER AMERICAN ASTROPHYSICIST, AND THE EARLY DEVELOPMENT OF AMERICAN ASTROPHYSICS
Cambridge: Cambridge University Press, 1984, ISBN 0-521-26582-7. 411p; illus. Bibl (3), index (3).

A scholarly biography of a now little known early astrophysicist (1857–1900), and the small group of pioneers which followed him.

442 Payne-Gaposchkin, Cecilia, (Katherine Haramundanis, ed.)
AN AUTOBIOGRAPHY AND OTHER RECOLLECTIONS
Cambridge: Cambridge University Press, 1984, ISBN 0-521-25752-2. 269p; illus. Bibl (x, 3), index (3).

The late Cecilia Payne-Gaposchkin was 'acclaimed in her own lifetime as the greatest woman astronomer of all time'. Her story, titled *The dyer's hand*, is prefaced by 67 pages of introductory matter, including a personal recollection by her daughter, the book's editor. Includes a chronological listing of Payne-Gaposchkin's writings.

443 Peltier, Leslie C.
STARLIGHT NIGHTS: THE ADVENTURES OF A STAR-GAZER
Cambridge, MA: Sky, 1980, ISBN 0-933346-02-6 pbk (reprint of 1st ed. 1965). 236p; illus.

The autobiography of an amateur astronomer, which conveys clearly his excitement and love for a subject which was much more than just a hobby.

444 Saward, Dudley
BERNARD LOVELL: A BIOGRAPHY
London: Robert Hale, 1984, ISBN 0-7090-1745-6. 320 + 12p; illus, diags. Index (2).

The authorized biography of the man who was responsible for the planning and building of the Jodrell Bank radio telescopes. Covers his life from school days to the present, ending with a summary of his views on cosmology. Includes a who's who of persons mentioned in the text. (*See also* 418).

445 van Woerden, Hugo, Willem N. Brouw and Henk C. van de Hulst
OORT AND THE UNIVERSE: A SKETCH OF OORT'S RESEARCH AND PERSON
Dordrecht: Reidel, 1980, ISBN 90-277-1180-1; 90-277-1209-3 pbk.

210p; illus, diags. Bibl (3), indexes (3).

A collection of 22 articles written to celebrate the 80th birthday, in 1979, of the renowned Dutch astronomer Jan H. Oort. Includes a sample manuscript by Oort, written in May 1980, and a full bibliography (with a subject index) of his publications, 1922–1979.

446 Webb, George Ernest
TREE RINGS AND TELESCOPES: THE SCIENTIFIC CAREER OF A.E. DOUGLASS
Tucson, AZ: University of Arizona Press, 1983, ISBN 0-8165-0798-8. 242p; illus. Bibl (3), index (2).

The biography of an individualistic Arizona astronomer (1867–1962), who turned to dendrochronology after his retirement in 1937.

447 Westfall, Richard S.
NEVER AT REST: A BIOGRAPHY OF ISAAC NEWTON
Cambridge: Cambridge University Press, 1980, ISBN 0-521-23143-4; 1983, ISBN 0-521-27435-4 pbk. 908p; illus, diags. Bibl (3), index (3).

A thoroughly researched account 'in which Newton's scientific career furnishes the central theme'. The author's goal was to 'present his science, not as the finished product which has done so much to shape the whole of the modern intellect, but as a developing endeavour of a living man confronting it as a problem still to be solved'.

Two earlier biographies are Frank E. Manuel, *A portrait of Isaac Newton* (Washington, DC: New Republic Books, 1968; London: Frederick Muller, 1980, ISBN 0-584-95357-7, 478p), and Colin A. Ronan, *Sir Isaac Newton* (London: International Textbook Co., 1969, ISBN 0-249-44007-5, 91p). (*See also* 433).

4 Astronomy Plus . . . (multi- and interdisciplinary works)

General

448 Clark, Cumberland
ASTRONOMY IN THE POETS
Philadelphia: R. West, 1982, ISBN 0-8492-0455-0 (reprint of 1st ed. 1922). 116p.

A literary study of astronomical references in Shakespeare and Milton's *Paradise lost*, with a long final chapter covering other poets.

449 Cronin, Vincent
THE VIEW FROM PLANET EARTH: MAN LOOKS AT THE COSMOS
London: Collins, 1981, ISBN 0-00-211397-X; New York: Morrow Quill, 1983, ISBN 0-688-00642-6; 0-688-01479-8 pbk. 348 + 12p; illus. Bibl, index (2).

A historical survey by a historian and biographer, which takes a multidisciplinary approach to the development of mankind's view of the nature of the universe. Attempts to bridge the gap between the 'two cultures' by showing the effect of science on art, and vice versa. Covers astronomy up to the present, and includes some discussion of science fiction and UFOs. Ends with an appendix by Nigel Henbest, "The probability of life: an astronomer's view".

450 Culver, Roger B. and Philip A. Ianna
THE GEMINI SYNDROME: STAR WARS OF THE OLDEST KIND
Tucson, AZ: Pachart, 1979, ISBN 0-912918-17-9. 216p; illus, diags. Bibl (2), index (2).

An investigation of astrology which concentrates on astronomical aspects.

451 Davidson, Norman
ASTRONOMY AND THE IMAGINATION: A NEW APPROACH TO MAN'S EXPERIENCE OF THE STARS
London: Routledge & Kegan Paul, 1985, ISBN 0-7102-0371-3. 237p; diags. Bibl (3), index (2).

In direct contrast to those books which painstakingly explain the true nature of celestial bodies and their movements, this book returns to first principles by examining the heavens as part of the ordinary person's wider environment, and considering the geocentric viewpoint as the true reflection of how the heavens are perceived by most people. The author presents observational astronomy as a part of nature and a creative science in its own right, influenced by mythology, literature, history, psychology and philosophy.

452 Grossinger, Richard
THE NIGHT SKY: THE SCIENCE AND ANTHROPOLOGY OF THE STARS AND PLANETS
San Francisco: Sierra Club Books, 1981, ISBN 0-87156-288-X. 481p. Bibl (3), index (3).

A wide-ranging book about the stars, the planets and their relationship to mankind. According to the author, it is about the attempt of our species to locate itself in the vast universe in which it has arisen. Part of the book tells the story of Western astronomy, another part is about science fiction. The author also discusses mythology, astrology and flying saucers as manifestations of the need to come to terms with the universe.

453 Hastings, Aryan W. and Jeanette Randall
ASTRONOMICAL CARTOONS
Indianapolis, IN: Mt. Olympus Press, 1982. 44p; illus.

A collection of humorous cartoons on astronomical and space themes.

454 Hawkins, Gerald S.
MINDSTEPS TO THE COSMOS
New York: Harper & Row, 1983, ISBN 0-06-015156-0. 340p; illus, diags. Bibl (1), index (2).

The author presents his theory of 'mindsteps': steps forward in our understanding of the cosmos. He proposes five steps (numbered 0–4), each beginning with a new idea and followed by a period of adjustment; for example, the final mindstep, number four (late 19th to early 20th century), when the idea of leaving the Earth and its atmosphere became feasible. Incorporates some new work from the field of archaeoastronomy, and includes a number of computer programs.

455 Hoyle, Fred
TEN FACES OF THE UNIVERSE
London: Heinemann, 1977, ISBN 0-435-54427-6; San Francisco: Freeman, 1977, ISBN 0-7167-0384-X; 0-7167-0383-1 pbk. 207p; illus, diags. Bibl (x), index (1).

Looks at the nature of the universe from several points of view, those of the theologian, the physicist, the mathematician, the astrophysicist, the geophysicist, the biologist and others, concluding with a chapter on the problems facing mankind on Earth.

456 Hughes, David
THE STAR OF BETHLEHEM MYSTERY (UK)/THE STAR OF BETHLEHEM: AN ASTRONOMER'S CONFIRMATION (US)
London: Dent, 1979, ISBN 0-460-04343-9; New York: Walker, 1979, ISBN 0-8027-0644-4; New York: Pocket Books, 1980, ISBN 0-671-83679-X pbk. 218 + 8p; illus, diags. Bibl (2), glos (1), index (2).

An 'investigation into the astronomical, biblical, historical, religious and astrological evidence about the star'. A thorough piece of detective work, which considers rival theories, but concludes that the most probable explanation is that the 'star' was actually a conjunction of Jupiter and Saturn.

457 Jackson, J. Lawrence
HUMANITY AND THE UNIVERSE
Bognor Regis: Hew Horizon, 1983, ISBN 0-86116-787-2. 227p; illus, diags.

An attempt to explain the nature of the universe, the Earth, and humanity's place in the cosmos. A personal account which includes speculation on UFOs and their origin.

458 Jastrow, Robert
GOD AND THE ASTRONOMERS
New York: Norton, 1978, ISBN 0-393-011870-9; New York: Warner, 1980, ISBN 0-446-97349-1 pbk. 136 + 8p; illus (some col.).

A brief account by an astronomer and self-professed agnostic of the religious implications of astronomical developments in the 20th century. Considers why some astronomers found it difficult to accept new cosmological theories, suggesting that they could not permit themselves the loss of the idea of a fixed beginning point.

459 Macvey, John W.
WHERE WILL WE GO WHEN THE SUN DIES?
Briarcliff Manor, NY: Stein & Day, 1983, ISBN 0-8128-2698-1. 264p; illus. Bibl (2), index (2).

Speculation on the possibility of extra-solar planets. Outlines why the other Solar worlds are not likely to be suitable for human life, then looks at different types of star and the kinds of planets they might support. Considers also the possibility of extraterrestrial life and cities in space.

460 Michanowsky, George
THE ONCE AND FUTURE STAR (rev. ed.)
New York: Barnes & Noble, 1979, ISBN 0-06-464027-2 pbk; London: Sphere, 1980, ISBN 0-7221-6070-4 pbk. 151p (UK 117 + 12p); illus, diags. Index (2).

A popular account of the discovery of the Vela pulsar (Vela X supernova remnant), and how it led the author in his researches to discover an ancient Sumerian tablet indicating that the supernova had been observed 6,000 years ago. He considers the effect of this event on subsequent theology and mythology. An epilogue added to this edition recounts the author's discovery of an Egyptian tradition associated with the Vela supernova.

461 Sagan, Carl, (produced by Jerome Agel)
THE COSMIC CONNECTION: AN EXTRATERRESTRIAL PERSPECTIVE
Garden City, NY: Doubleday (Anchor Press), 1980, ISBN 0-385-17365-2; London: Macmillan (Papermac), 1981, ISBN 0-333-32474-9 pbk (reprint of 1st ed. 1973). 288p; illus, diags. Index (2).

One of the world's best known science popularizers presents his personal view of the universe and our place in it. The book is divided into three parts: the first puts the Earth in its place in the vastness of the cosmos, the second discusses various aspects of the Solar System, while the third is devoted to the possibility of communication with extraterrestrial intelligence. Throughout this collection of essays the author speculates freely on matters where few, if any, facts are available.

462 Sagan, Carl
COSMOS
New York: Random House, 1980, ISBN 0-394-50294-9; 0-394-32597-4

pbk; 0-394-32687-3 pbk; London: Macdonald, 1981, ISBN 0-354-04531-8; 1984, ISBN 0-354-09722-6 pbk; London: Futura, 1983, ISBN 0-7088-1996-6 pbk; New York: Ballantine, 1985, ISBN 0-345-30005-X pbk. 365p; illus (mostly col.), diags. Bibl (2), index (3).

Based on the successful TV series, the book's 13 chapters are intended to complement the 13 programmes. A fundamentally optimistic work, Sagan's personal voyage through the cosmos is an exploration of the development and extent of our knowledge, 'with a broad human perspective'. In Sagan's view, 'the deepest scientific questions excite the interests and passions of enormous numbers of people'. The book allows more detailed treatment of topics than was possible in the TV series, but loses none of the author's infectious enthusiasm. Note that the many excellent photographs and imaginative artwork are not reproduced in the mass market paperback editions.

Also produced were two accompanying volumes by A. Elwood and L. Wood, *A reader/study guide for Cosmos* (New York: Random House, 1980, pbk, 278p), a collection of readings, and *A viewer's guide to Cosmos* (Random House, 1980, pbk, 71p).

463 Solomon, Joan
SPACE, COSMOLOGY AND FICTION
Oxford: Blackwell, 1985, ISBN 0-631-91030-1 pbk. 47p; illus, diags.

Part of the 'Science in a social context' (SISCON) series, aimed primarily at pre-college students, this volume looks briefly at early views of the universe, the development of modern cosmology, science fiction becoming fact in the form of space travel, and the uses of space.

Cosmic Evolution

The books listed here are those which look at the universe and the place of human life in it from the broadest perspective, describing its evolution from the beginning (the Big Bang), through the formation of the galaxies, stars, planetary systems, particularly the Solar System, the Earth, the origin of life, its evolution and development, and, in some cases, considering likely future events to the end of the universe. (*See also* 81).

464 Asimov, Isaac
THE EXPLODING SUNS: THE SECRETS OF THE SUPERNOVAS
New York: Dutton (Truman Talley), 1985, ISBN 0-525-24323-2;

London: Michael Joseph, 1985, ISBN 0-7181-2630-0. 276p; illus. Index (2).

An account of life on Earth and its link with supernovae, listed in this section because of its wide-ranging coverage of background events. Begins with an account of the development of astronomical knowledge, from the belief in an unchanging sky to the discovery of celestial explosions: supernovae and the Big Bang. Then describes the make-up of the universe, from element formation to the creation of stars, planets and life, concluding with a look at the future of life and the possible effect of material ejected in supernova explosions on evolution.

465 Chaisson, Eric
COSMIC DAWN: THE ORIGINS OF MATTER AND LIFE
Boston: Little, Brown, 1981, ISBN 0-316-13590-9. 302p; illus, Bibl (1).

About half of this book is devoted to physical evolution, while the other half deals with biological and social evolution. It is intended for 'eclectic individuals having an interest in nature'. Unfortunately it lacks an index.

466 Cloud, Preston
COSMOS, EARTH, AND MAN: A SHORT HISTORY OF THE UNIVERSE
New Haven and London: Yale University Press, 1978, ISBN 0-300-02146-1. 372p; illus, diags. Bibl (x), index (2).

Written after the author's long, pioneering career in biogeology, this account strives to avoid oversimplification. It is intended for the 'educated layman'.

467 Gribbin, John
GENESIS: THE ORIGINS OF MAN AND THE UNIVERSE
London: Dent, 1981 ISBN 0-460-04505-9; New York: Delacorte, 1981, ISBN 0-440-02832-9; Oxford: Oxford University Press, 1982, ISBN 0-19-283035-X pbk; New York: Dell (Delta), 1982, ISBN 0-385-28321-0 pbk. 360p; illus, diags. Bibl (2), index (3).

This book encompasses knowledge from many fields of study in the author's quest to answer the questions, 'Where did we come from and where are we going?' Two long appendices discuss the age of the universe, and growth and limits.

468 Henbest, Nigel
THE EXPLODING UNIVERSE
London: Marshall Cavendish, 1979, ISBN 0-85685-509-X; New York: Macmillan, 1979, ISBN 0-02-550920-9. 217p; illus (mostly col.).

Glos (2), bibl (1), index (2). Q

A multidisciplinary account of the known universe, but one which differs from most of the other books in this section in that it begins with the Earth and moves outward.

469 Jastrow, Robert
RED GIANTS AND WHITE DWARFS (2nd ed.)
New York: Norton, 1979, ISBN 0-393-85002-1; New York: Warner, 1980, ISBN 0-446-97349-1 pbk. 275 + 24p; illus (some col.), diags. Bibl (1), index (2).

This book should be considered in conjunction with its sequel (470). It originated in a series of 57 half-hour CBS TV programmes in 1964, and provides a very good introduction to events in the universe as far as the threshold of human existence.

470 Jastrow, Robert
UNTIL THE SUN DIES
New York: Norton, 1977, ISBN 0-393-06415-8; London: Souvenir Press, 1978, ISBN 0-285-62338-9; Glasgow: Fontana/Collins, 1979, ISBN 0-00-635258-8 pbk; New York: Warner, 1980, ISBN 0-446-32195-8 pbk. 172p; illus. Index (2).

The sequel to (359), this volume tells the story of life from beginning to likely end. A mixture of fact and speculation firmly based on scientific research and discovery. Also discusses the possibility of extraterrestrial life.

471 Kandel, Robert S.
EARTH AND COSMOS: A BOOK RELATING THE ENVIRONMENT OF MAN ON EARTH TO THE ENVIRONMENT OF EARTH IN THE COSMOS
Oxford: Pergamon Press, 1980, ISBN 0-08-025061-5; 0-08-023086-5 pbk. 254p; illus, diags. Bibl (3), indexes (2).

The author, an astronomer, attempts to do what the rather cumbersome subtitle promises in this largely non-mathematical, but occasionally technical account, which is probably the most demanding of those listed here.

472 Pellegrino, Charles R. and Jesse A. Stoff
DARWIN'S UNIVERSE: ORIGINS AND CRISES IN THE HISTORY OF LIFE
New York: Van Nostrand, 1983, ISBN 0-442-27526-9. 208p; illus, diags. Bibl (3), index (2).

This book is concerned primarily with the origin and development of life on Earth, but it also provides an account of the preceding events in the universe. The bibliography contains mostly technical references.

473 Reeves, Hubert, (Ruth A. and John S. Lewis, trans.)
ATOMS OF SILENCE: AN EXPLORATION OF COSMIC EVOLUTION
Cambridge, MA: MIT Press, 1984, ISBN 0-262-18112-6 (trans. of 1st French ed. 1981). 244p; illus, diags. Bibl (1), index (2).
A non-technical account of cosmic evolution and physical processes in the universe, written in an appealing personal style.

474 Seielstad, George A.
COSMIC ECOLOGY: THE VIEW FROM THE OUTSIDE IN
Berkeley, CA: University of California Press, 1983, ISBN 0-520-04753-2. 169p; illus (some col.), diags. Bibl (x), index (2).
A concise account focusing on processes of evolution: atomic, chemical, biological, human and cultural. The text is brief, but the author provides a good guide to further reading.

Life in the Universe

General

475 Asimov, Isaac
EXTRATERRESTRIAL CIVILIZATIONS
New York: Crown, 1979, ISBN 0-517-53079-9; New York: Fawcett (Columbine), 1980, ISBN 0-449-90020-7 pbk; London: Robson, 1980, ISBN 0-86051-096-4; London: Pan, 1981, ISBN 0-330-26249-1 pbk. 316p. Index (2).
A non-technical, but thorough step by step presentation of the case for the existence of extraterrestrial civilizations. Necessarily speculative, but logically argued.

*476 Baugher, Joseph F.
ON CIVILIZED STARS
Englewood Cliffs, NJ: Prentice-Hall, 1985, ISBN 0-13-634429-1; 0-13-634411-9 pbk. 260p.
This book examines all aspects of life in the universe: the formation of the Solar System, biological evolution on Earth, the likelihood of other civilizations in our galaxy, and how to communicate with them. (From *S&T* Sept. 1985, p 230).

477 Billingham, John (ed.)
LIFE IN THE UNIVERSE
Cambridge, MA: MIT Press, 1981, ISBN 0-262-02155-2; 0-262-52062-1 pbk. 461p; illus, diags. Glos (3), index (2).

The proceedings of a conference held at the NASA Ames Research Center in June 1979 presents speculation based on currently accepted scientific knowledge in diverse fields. Often technical, but an interesting corrective to the simplistic ideas derived from bad science fiction, particularly that of TV and the cinema. The book's four parts cover the origin of life, life-supporting environments, the evolution of complex life in the galaxy, and the detectability of technological civilizations.

478 Boyce, Chris
EXTRATERRESTRIAL ENCOUNTER: A PERSONAL PERSPECTIVE
Newton Abbot: David & Charles, 1979, ISBN 0-7153-7634-9; London: New English Library, 1981, ISBN 0-450-05009-2 pbk. 199 + 8p; illus. Bibl (2), index (1).

A readable, almost chatty book by an author who has written science fiction, based on the proposition that extraterrestrial life is likely to be far stranger than we have so far imagined. Attempts to show that most other ideas on the subject are hopelessly naive.

479 Bracewell, Ronald
THE GALACTIC CLUB: INTELLIGENT LIFE IN OUTER SPACE
London: Heinemann, 1978, ISBN 0-434-08520-0 pbk; New York: Norton, 1979, ISBN 0-393-95022-0 pbk (reprint of 1st ed. 1974), 143p; illus, diags. Bibl (2), index (2).

A brief survey of speculation on the possibility of extraterrestrial life and the likelihood of making contact with it. Takes a historical perspective, distilling 'the essential simplicity of the new thoughts as they have come along' to free them of mathematics and scientific jargon.

480 Breuer, Reinhard, (Cecilia Payne-Gaposchkin and Mark Lowery, trans.)
CONTACT WITH THE STARS: THE SEARCH FOR EXTRATERRESTRIAL LIFE
Oxford and San Francisco: Freeman, 1982, ISBN 0-7167-1355-1; 0-7167-1389-6 pbk (trans. of 1st German ed. 1978). 292p; illus, diags. Bibl (2), index (2).

A scholarly survey of speculation on the possible existence of extraterrestrial intelligence, and possible means of making contact with it, which should be accessible to the interested general reader. The English translation has a new preface by the author written in November 1980.

481 Bylinsky, Gene
LIFE IN DARWIN'S UNIVERSE: EVOLUTION AND THE COSMOS
Garden City, NY: Doubleday, 1981, ISBN 0-385-17049-1. 238 + 8p; illus (some col.), diags. Index (2).

Following a brief account of the beginning and subsequent evolution of the universe up to the development of life on Earth, the author speculates on the kinds of life that may have arisen in various different environments. Finally, he discusses the possibility of locating life-bearing planets and establishing contact with their inhabitants.

482 Christian, James L. (ed.)
EXTRATERRESTRIAL INTELLIGENCE: THE FIRST ENCOUNTER
Buffalo, NY: Prometheus Books, 1976, ISBN 0-87975-063-4. 303p; illus, diags. Bibl (2).

Fifteen speculative essays (one a reprint) by writers from various fields, including several philosophers, and representatives of astronomy, religion, science fiction and television (Leonard Nimoy of *Star Trek*, who has played the part of an intelligent extraterrestrial).

483 Delano, Kenneth J.
MANY WORLDS, ONE GOD
Hicksville, NY: Exposition, 1977, ISBN 0-682-48644-2. 126p. Bibl (2).

A Christian view of the possibility of the existence of extraterrestrial civilizations. Discusses UFOs, and the religious implications of contact. Promotes the view of God as 'father of a great multitude of intelligent species living on worlds both near and far from us'.

484 Dick, Steven J.
PLURALITY OF WORLDS: THE ORIGINS OF THE EXTRATERRESTRIAL LIFE DEBATE FROM DEMOCRITUS TO KANT
Cambridge: Cambridge University Press, 1982, ISBN 0-521-24308-4; 1984, 0-521-31985-4 pbk. 246p; illus, diags. Bibl (3), index (3).

This scholarly work goes well beyond the bounds of astronomy to encompass the history of science as a whole, pointing out that 'the plurality of worlds concept is demonstrably enmeshed in many of the great scientific traditions, notably those centred around Aristotle, the ancient atomists, Copernicus, Descartes, and Newton'.

485 Edelson, Edward
WHO GOES THERE? THE SEARCH FOR INTELLIGENT LIFE IN THE UNIVERSE
New York: Doubleday, 1979, ISBN 0-385-11257-2; New York: McGraw-Hill, 1980, ISBN 0-07-018986-2; London: New English Library, 1980, ISBN 0-450-04618-4 pbk. 228p; diags. Index (2).

A popular account which uses the Viking mission to Mars as a take-off point.

486 Feinberg, Gerald and Robert Shapiro
LIFE BEYOND EARTH: THE INTELLIGENT EARTHLING'S GUIDE TO LIFE IN THE UNIVERSE
New York: Morrow, 1980, ISBN 0-688-03642-2. 464p; illus, diags. Bibl (3), index (2).

A more serious work than the title suggests, this account deals with life on Earth, its origin and evolution; life in general and the conditions necessary for its existence; life in different environments; and the possibility of detecting extraterrestrial life.

487 Goldsmith, Donald (ed.)
THE QUEST FOR EXTRATERRESTRIAL LIFE: A BOOK OF READINGS
Mill Valley, CA: University Science Books, 1980, ISBN 0-935702-08-3; 0-935702-02-4 pbk. 308p; illus, diags. Bibl (3), index (1). Q

Contains 58 contributions spanning 2,000 years, including Lucretius, Giordano Bruno, Christiaan Huygens, Svante Arrhenius, Fred Hoyle and Chandra Wickramasinghe, Percival Lowell, Freeman Dyson, Carl Sagan and Gerard O'Neill. Too technical for the general reader in parts, but overall a good survey of a subject on the fringe of conventional science.

488 Goldsmith, Donald and Owen Tobias
THE SEARCH FOR LIFE IN THE UNIVERSE
Menlo Park, CA: Benjamin/Cummins, 1980, ISBN 0-8053-3325-8 pbk. 436p; illus, diags. Bibl (x), index (1).

This is virtually a general astronomy textbook for the non-science undergraduate or lay reader, but is structured round the concept of finding life elsewhere in the universe. The book's five parts cover: 1. why do we search? 2. the universe at large, 3. life, 4. the search for life in the Solar System, and 5. the search for extraterrestrial intelligence. Includes questions.

489 Hart, Michael H. and Ben Zuckerman (ed.)
EXTRATERRESTRIALS: WHERE ARE THEY?
New York: Pergamon Press, 1982, ISBN 0-08-026342-9; 0-08-026341-0. 182p; illus, diags. Bibl (x), index (2).

This collection of 17 papers from a symposium held at the University of Maryland in 1979 is often quite technical, requiring some mathematical knowledge on the part of the reader. Covers a wide range of topics, including interstellar travel and the colonization of other worlds.

490 Morrison, Philip, John Billingham and John Wolfe (ed.)
THE SEARCH FOR EXTRATERRESTRIAL INTELLIGENCE: SETI
Washington, DC: NASA, 1977, NASA SP-419; New York: Dover, 1979, ISBN 0-486-23890-3 pbk. 176p; diags. Bibl (1).

This report of NASA workshops is technical in places, and will likely only interest the reader with a strong interest in SETI.

*491 Poynter, Margaret and Michael J. Klein
COSMIC QUEST
New York: Atheneum, 1984, ISBN 0-689-31068-4. 124p; illus. Glos, bibl.

A compact but substantive introduction to the search for life elsewhere. Discusses how life may have been formed, where it might now be, and how we might look for it. Well suited to both pre-college students and general readers of all ages. (From *S&T* April 1985, p 325).

492 Regis, Edward, jr. (ed.)
EXTRATERRESTRIALS: SCIENCE AND ALIEN INTELLIGENCE
Cambridge: Cambridge University Press, 1985, ISBN 0-521-26227-5. 278p. Bibl (x), index (2).

This collection of scholarly essays, which should suit the seriously interested general reader, considers such topics as the possible existence of extraterrestrials, how they may have evolved and whether they might be intelligent, their modes of understanding and whether they are compatible with ours, how we might communicate, and their moral codes.

493 Ridpath, Ian
LIFE OFF EARTH
London: Granada, 1983, ISBN 0-246-11933-0. 128p; illus (mostly col.), diags. Index (2).

A popular account of all aspects of life beyond Earth's atmosphere, including not only the possibility of extraterrestrial intelligence, but also human activity in space. Covers such topics as the Space Shuttle, colonies in space, and possible visits by alien beings to Earth.

494 Ridpath, Ian
SIGNS OF LIFE: THE SEARCH FOR LIFE IN SPACE
Harmondsworth: Penguin (Kestrel), 1977, ISBN 0-7226-5275-5. 190p; illus, diags. Bibl (1), index (2).

This non-technical account is aimed primarily at younger readers, but provides a good general summary of our knowledge of the origin of life on Earth, evolution, the likely future of mankind, the origin of life in the Solar System, as well as a survey of attempts to communicate with extraterrestrial intelligences, and 'probes from the skies' (UFOs). Illustrated with photographs and cartoons. This is a revised version of *Worlds beyond* (1975).

495 Ridpath, Ian
MESSAGES FROM THE STARS
Glasgow: Fontana/Collins, 1978, ISBN 0-00-634596-4 pbk; New York: Harper & Row, 1978, ISBN 0-06-013589-1; 0-06-090734-7 pbk. 240p; diags. Bibl (x).

A sequel to (494) which seems to be aimed at a wider readership. Considers the basis for belief in the existence of extraterrestrial intelligence; the search for, and likely consequences of contact with such beings; and the possibility that we have already been visited.

496 Rood, Robert T. and James S. Trefil
ARE WE ALONE? THE POSSIBILITY OF EXTRATERRESTRIAL CIVILIZATIONS
New York: Scribner's, 1981, ISBN 0-684-16826-X; 1983, ISBN 0-684-17842-7 pbk. 262p; illus, diags. Index (2).

In writing this book the authors hope to fill what they see as 'an enormous gap in the perception of the problem of extraterrestrial intelligence by the public and by the scientific community'. They are far less optimistic than most other writers on this subject. Following an analysis of the knowledge which forms the foundations for the belief in the existence of extraterrestrial intelligence, they present their separate, sceptical conclusions.

497 Sagan, Carl (ed.)
COMMUNICATION WITH EXTRATERRESTRIAL INTELLIGENCE (CETI)
Cambridge, MA: MIT Press, 1973, ISBN 0-262-19106-7; 0-262-69037-3 pbk. 428p; illus, diags. Bibl (x), glos (2), indexes (3).

The proceedings of the First International Conference on Extraterrestrial Civilizations and the Problems of Contact with Them, held in September 1971 in Armenia: a joint US-Soviet venture. This 'verbatim record of an interdisciplinary scientific discussion' may not be easy for the lay reader to follow, but the wide-ranging exchange of ideas can be followed by any interested reader. Includes an elementary glossary for the novice.

498 Shklovskii, I.S. and Carl Sagan, (Paula Fern, trans.)
INTELLIGENT LIFE IN THE UNIVERSE
London: Pan (Picador), 1977, ISBN 0-330-25125-2 pbk; San Francisco: Holden Day, 1978, ISBN 0-8162-7913-6 pbk (reprint of 1st ed. 1966). 309p; illus (some col.), diags. Bibl (2), indexes (3).

Much expanded from Shklovskii's original Russian text of 1963, this pioneering work is intended as a semi-popular treatment of a subject on the fringe of conventional science. The book's three parts provide substantive treatment of 1. the evolution of the universe, from the Big Bang to the formation of the Solar System, 2. life: what it is, its existence on Earth, and its possible existence elsewhere, and 3. intelligent life and the possibility of contact.

Origin of life on Earth

All the books in this section are by Fred Hoyle (most with Chandra Wickramasinghe), presenting his controversial ideas about the extraterrestrial origin of life on Earth. They are listed in approximately chronological order, culminating with Hoyle's personal view of an intelligence-based cosmology.

499 Hoyle, Fred and Chandra Wickramasinghe
LIFECLOUD: THE ORIGIN OF LIFE IN THE UNIVERSE
London: Dent, 1978, ISBN 0-460-04335-8; London: Sphere, 1979, ISBN 0-7221-4755-4 pbk; New York: Harper & Row, 1979, ISBN 0-06-011954-3. 189 + 8p; illus, diags. Index (1).

Speculation on the astronomical basis of the origin of life. The authors argue that life on Earth arrived in showers of already living cells from comet-type bodies. Includes introductory chapters on terrestrial biology and astronomy. Also discusses the possibility of life elsewhere in the Solar System, and in other star systems. The ideas presented here and in the subsequent books have, on the whole, been rejected or ignored by other scientists.

500 Hoyle, Fred and Chandra Wickramasinghe
DISEASES FROM SPACE
London: Dent, 1979, ISBN 0-460-04357-9; New York: Harper & Row, 1979, ISBN 0-06-011937-3; London: Sphere, 1981, ISBN 0-7221-4754-6 pbk. 196p; diags. Bibl (2), index (1).

A sequel to (499) which develops the authors' theory that infectious diseases of plants and animals are caused by viruses and bacteria carried to this planet by comets.

501 Hoyle, Fred
THE RELATIONSHIP OF BIOLOGY TO ASTRONOMY
Cardiff: University College Cardiff Press, 1980, ISBN 0-906449-20-0 pbk. 25p; illus, diags. Bibl (1).

A semi-popular pamphlet which explores the hypothesis that 'interstellar grains are bacteria'.

502 Hoyle, Fred and Chandra Wickramasinghe
EVOLUTION FROM SPACE (UK)/EVOLUTION FROM SPACE: A THEORY OF COSMIC CREATIONISM (US)
London: Dent, 1981, ISBN 0-460-04535-0; New York: Simon & Schuster, 1982, ISBN 0-671-45031-X; 1984, ISBN 0-671-49263-2 pbk; St. Albans: Granada (Paladin), 1983, ISBN 0-586-08421-5 pbk. 176p; diags. Index (1).

In the sequel to (499) and (500) the authors argue that not only did life on Earth originate in space, but that its continuing evolution has been influenced by the arrival of new genes from space. The complexity of life on Earth, they claim, could not be a result of random processes. They consider the possibility of a cosmic intelligence, and the ultimate origin of life/intelligence (*see* 506). Includes technical references.

503 Hoyle, Fred
EVOLUTION FROM SPACE (THE OMNI LECTURE) (UK)/ EVOLUTION FROM SPACE (THE OMNI LECTURE) AND OTHER PAPERS ON THE ORIGIN OF LIFE (US)
Cardiff: University College Cardiff Press, 1982, ISBN 0-906449-38-3 pbk; Hillside, NJ: Enslow, 1982, ISBN 0-89490-083-8. 30p (UK), 80p (US).

The US edition contains three papers, one by Chandra Wickramasinghe, introducing his and Hoyle's ideas to a general readership.

504 Hoyle, Fred and Chandra Wickramasinghe, (Barbara Hoyle, ed.)
SPACE TRAVELLERS: THE BRINGERS OF LIFE
Cardiff: University College Cardiff Press, 1981, ISBN 0-906449-27-8. 197p; illus, diags. Index (2).

Bacteria as interstellar travellers: part of the authors' overall thesis that life on earth had a cosmic origin. They discuss evidence which suggests that space-travelling bacteria caused major influenza epidemics in history, and suggest that bacteria play a role in adapting planets to their physical needs, and possibly also influence star formation. An appendix provides "A technical note on influenza".

505 Hoyle, Fred and Chandra Wickramasinghe
LIVING COMETS
Cardiff: University College Cardiff Press, 1985, ISBN 0-906449-79-0 pbk. 133p; illus, diags. Bibl (1), index (1).

A fairly technical book in which the authors develop their thesis of comets as bearers of organic matter, and their theory that comets carry living micro-organisms to Earth.

506 Hoyle, Fred
THE INTELLIGENT UNIVERSE
London: Michael Joseph, 1983, ISBN 0-7181-2298-4; 1985, ISBN 0-7181-2613-0 pbk; New York: Holt, 1984, ISBN 0-03-070083-3. 256p; illus & diags (mostly col.). Index (2).

A popular re-examination of theories of the origin of the universe, with particular emphasis on the origin of life: its development and spread, the extraterrestrial origin of life on Earth and the development of intelligence. Argues against Darwin, adherence to whose views, Hoyle believes, is leading mankind to ultimate self-destruction. He also argues that there is a component of intelligent design missing from cosmological studies.

5 Astrophysics

Related Physics

507 Calder, Nigel
EINSTEIN'S UNIVERSE
London: BBC, 1979, ISBN 0-563-17592-3; New York: Viking, 1979, ISBN 0-670-29076-9; Harmondsworth: Penguin (Pelican), 1982, ISBN 0-14-005499-5 pbk. 178p; illus, diags. Index (2).

Published in conjunction with a special TV programme to mark the centenary of Einstein's birth, this book was separately conceived and written 'to convey the ideas and experiments in printed words rather than with moving pictures'. Aims to make relativity clear to the lay reader.

508 Calder, Nigel
THE KEY TO THE UNIVERSE
London: BBC: 1977, ISBN 0-563-17091-3; New York: Viking, 1977, ISBN 0-670-41270-8; Harmondsworth: Penguin, 1978, ISBN 0-14-005065-5 pbk. 196p; illus, diags. Bibl (1), index (2). Q

Written to accompany the TV programme first broadcast by the BBC in January 1977, this book introduces the general reader to the esoteric field of particle physics and its implications for astrophysics and cosmology.

Another account, but with less astronomical content, is James S. Trefil's *From atoms to quarks* (New York: Scribner's, 1980, ISBN 0-684-17460-X; London: Athlone Press, 1980, ISBN 0-485-11204-3, 225p).

509 Davies, Paul
THE EDGE OF INFINITY: NAKED SINGULARITIES AND THE DESTRUCTION OF SPACETIME
London: Dent, 1981, ISBN 0-460-04490-7; New York: Simon & Schuster, 1982, ISBN 0-671-44063-2; 1983, ISBN 0-671-44062-5 pbk; Oxford: Oxford University Press, 1983, ISBN 0-19-286031-3 pbk. 194p; diags. Index (2).

An account of singularities, areas of infinite gravity where known physical laws are suspended, intended for readers without a background knowledge of mathematics or physics.

510 Davies, Paul
OTHER WORLDS
London: Dent, 1980, ISBN 0-460-04400-1; New York: Simon & Schuster, 1981, ISBN 0-671-42227-8; London: Sphere (Abacus), 1982, ISBN 0-349-10741-6 pbk. 207p; diags. Index (2).

An introduction to quantum theory, which discusses the importance of sub-atomic physics to cosmological studies. Considers also the broader issues: the nature of reality, what is man?, and is the universe a random accident or the outcome of a delicate selection process? Discusses the anthropic principle: the idea that the existence of mankind is not an accident, but the natural outcome of some evolutionary process at work in the universe.

See also the author's *God and the new physics* (London: Dent, 1983, ISBN 0-460-04577-6; New York: Simon & Schuster, 1983, ISBN 0-671-47688-2, 255p).

511 Davies, P.C.W. [Paul]
THE SEARCH FOR GRAVITY WAVES
Cambridge: Cambridge University Press, 1980, ISBN 0-521-23197-3. 144p; illus, diags. Index (2).

The story of an aspect of research in physics which has implications for astronomy. The author foresees the invention of a 'gravity telescope' which could be used to study quasars and neutron stars, and to probe the edges of black holes. Intended for non-specialists, the book uses elementary algebra and calculus. A mathematical appendix deals with "The theory of gravity wave detectors".

512 Davies, P.C.W. [Paul]
SPACE AND TIME IN THE MODERN UNIVERSE
Cambridge: Cambridge University Press, 1977, ISBN 0-521-21445-9; 0-521-29151-8 pbk. 232p; diags. Index (2).

A review of recent developments in our knowledge of the space-time structure, which also shows how views of the universe have changed since the time of Newton. Many of the ideas discussed overlap the realms of religion and metaphysics. Assumes little expertise on the part of the reader, but the author suggests that, because much of the material is very advanced, students of physics, astronomy, applied mathematics and philosophy will find the book useful for background reading.

513 Geroch, Robert
GENERAL RELATIVITY FROM A TO B
Chicago and London: University of Chicago Press, 1978, ISBN 0-226-28863-3. 225p; diags. Index (1).

This non-technical introduction, less demanding than Lilley (516), begins with a section on the historical background to Einstein's work, showing why earlier ideas were unrealistic.

514 Gribbin, John
SPACEWARPS
New York: Delacorte (Eleanor Friede), 1983, ISBN 0-385-29266-X; 1984, ISBN 0-385-29366-6 pbk; Harmondsworth: Penguin (Pelican), 1984, ISBN 0-14-022531-5 pbk. 211p; illus, diags. Bibl (2), index (3).

A companion to the author's *Timewarps* (Delacorte; Dent, 1979), which, like the earlier book, also deals with bent spacetime, but with the emphasis on space. Discusses the implications of Einstein's theory of relativity on the universe as we know it, covering phenomena such as black holes, white holes, quasars, neutrinos, time travel, and life, as well as what lies 'beyond Einstein'.

515 Kaufmann, William J., III
RELATIVITY AND COSMOLOGY (2nd ed.)
New York: Harper & Row, 1977, ISBN 0-06-043572-0 pbk. 154p; illus, diags. Glos (2), index (1).

A guide for the general reader to the 'more exotic and fascinating aspects of current research, particularly general relativity'. This well presented account uses many illustrations and line drawings to explain difficult, often still speculative concepts.

516 Lilley, Sam
DISCOVERING RELATIVITY FOR YOURSELF
Cambridge: Cambridge University Press, 1981, ISBN 0-521-23038-1; 0-521-29780-X pbk. 425p; diags. Index (2).

A book intended for the determined lay reader, written not by an expert on relativity, but by an 'expert . . . in the art of teaching science to the non-scientist'. Uses a simplified version of the infinitesimal calculus ('*the* calculus of school terminology'), but this is introduced gradually.

517 Narlikar, Jayant V.
THE LIGHTER SIDE OF GRAVITY
San Francisco: Freeman, 1982, ISBN 0-7167-1343-8; 0-7167-1344-6 pbk. 194p; illus, diags. Index (2).

A popular discussion of the least well understood of the basic forces of nature ('lighter' is used to mean non-technical, not non-serious). Deals with the theory of gravitation from Newton to Einstein, then covers modern developments, particularly those concerning black holes. Focuses on astronomical phenomena, where the effects of gravity are more readily seen and studied.

518 Nicolson, Iain
GRAVITY, BLACK HOLES AND THE UNIVERSE
Newton Abbot: David & Charles, 1981, ISBN 0-7153-7849-X; New York: Halsted Press, 1981, ISBN 0-470-27111-6. 264p; illus, diags. Index (2).

In this book for the seriously interested lay reader the author attempts, in the first part, 'to show how our ideas on force, motion and gravity developed from earliest times to the triumphant syntheses achieved by Newton and Einstein'. He goes on to discuss the theoretical and observational aspects of black holes and cosmology, where gravity plays a significant role. Finally he looks at the present and future of general relativity, and alternative theories of gravitation.

519 Wald, Robert M.
SPACE, TIME, AND GRAVITY: THE THEORY OF THE BIG BANG AND BLACK HOLES
Chicago and London: University of Chicago Press, 1977, ISBN 0-226-87030-8. 131p; illus, diags. Bibl (1), index (1).

Based on a series of nine lectures given in 1976, this book covers Einstein's treatment of gravity, cosmology and the evolution of the universe. The first five chapters focus on stellar evolution and gravitational collapse, while the last four concentrate on black holes. Intended for the general reader, the book might also serve as an introduction for students. Uses some algebra.

Beyond the visible

520 Adams, D.J.
COSMIC X-RAY ASTRONOMY
Bristol: Adam Hilger, 1980, ISBN 0-85274-253-3. 150p; illus, diags. Bibl (2), index (2).

This book, intended for the non-specialist with some knowledge of astronomy and physics, reviews the progress made in x-ray astronomy from 1962–78. Does not include results from the Einstein observatory (HEAO-2) which was launched in late 1978. Includes technical references.

521 Culhane, J. Leonard and Peter W. Sanford
X-RAY ASTRONOMY
London: Faber, 1981, ISBN 0-571-11550-0; New York: Scribner's, 1981, ISBN 0-684-16725-5. 192 + 16p; illus, diags. Bibl (1), index (2).

A general introduction to this relatively new branch of astronomy, which can be carried out only by means of instruments taken above about 30km from the Earth's surface. Technical in places, but mostly accessible to the reader with some previous astronomical knowledge.

522 Field, George B. and Eric J. Chaisson
THE INVISIBLE UNIVERSE: PROBING THE FRONTIERS OF ASTROPHYSICS
Boston: Birkhauser, 1985, ISBN 0-8176-3235-2. 195 + 20p; illus (some col.), diags. Glos (3), bibl (1), index (2).

A survey of new results from astronomy outside the visible part of the electromagnetic spectrum, derived from the three-volume report edited by George B. Field, *Astronomy and astrophysics for the 1980s* (Washington, DC: National Academy Press, 1982–84). Provides broad coverage of astrophysical research for the general reader. Includes a discussion of the search for life elsewhere in the universe.

523 Hey, J.S.
THE RADIO UNIVERSE (3rd ed.)
Oxford: Pergamon Press, 1983, ISBN 0-08-029152-X; 0-08-029151-1 pbk. 246p; illus, diags. Index (1).

This book by a pioneer in the field of radio astronomy is 'designed for the intelligent layman', but 'should also provide a useful survey for those scientists who look for an easily readable outline of progress in radio astronomy and its contribution to our understanding of the universe'. Covers history, the properties of radio waves, radio telescopes, research in radio astronomy, and radar astronomy.

For a less technical introduction *see* Gerrit L. Verschuur, *The invisible universe: the story of radio astronomy* (New York: Springer-Verlag, 1974, ISBN 0-387-90078-0; London: The English Universities Press, 1974, ISBN 0-340-19180-5 pbk, 173p).

524 Marten, Michael and John Chesterman
THE RADIANT UNIVERSE: ELECTRONIC IMAGES FROM SPACE
Poole: Blandford, 1980, ISBN 0-7137-1116-7; New York: Macmillan, 1980, ISBN 0-02-580420-0. 128p; illus (mostly col.). Index (1). Q

A picture book featuring spectacular images of a wide range of celestial objects seen outside the visible part of the electromagnetic spectrum. False colours have been added to make the images visible to the human eye.

525 Tucker, Wallace H.
THE STAR SPLITTERS: THE HIGH ENERGY ASTRONOMY OBSERVATORIES
Washington, DC: NASA, 1984, NASA SP-466. 182p; illus (mostly col.), diags. Bibl (1).

This non-technical account of the NASA HEAO programme is divided into two parts: the first tells the story of the development and construction of the observatories, and the people involved; the second, slightly shorter, part is a survey of the results obtained by the three HEAO satellites, which were designed to investigate x-ray, gamma ray and cosmic ray sources.

526 Worvill, Roy
THE RADIO UNIVERSE: AN INTRODUCTION TO RADIO ASTRONOMY AND OUTER SPACE
London: Kahn & Averill, 1977, ISBN 0-900707-42-9; New York: Orbiting Book Service, 1977, ISBN 0-914326-10-4. 96p; diags. Glos (2), bibl (1).

A brief, readable introduction which assumes no previous knowledge of astronomy.

Cosmology

General/history

527 Blacker, Carmen and Michael Loewe (ed.)
ANCIENT COSMOLOGIES
London: Allen & Unwin, 1975, ISBN 0-04-100038-2; Totowa, NJ: Rowman & Littlefield, 1975, ISBN 0-87471-683-7. 270 + 8p; illus, diags. Bibl (x), index (2).

A collection of nine scholarly essays, all but one based on lectures given at Cambridge University in 1972, which seek to answer the question, 'what was the shape of the universe imagined by those ancient peoples to whom all modern knowledge was inaccessible?' Topics covered are ancient Egypt, Sumer and Babylon, Jewish cosmology, early China, ancient India, Islamic cosmology, Scandinavia, Greek cosmologies, and the European heritage from the ancient systems. Not listed by Seal (24).

528 Durham, Frank and Robert D. Purrington
FRAME OF THE UNIVERSE: A HISTORY OF PHYSICAL COSMOLOGY
New York: Columbia University Press, 1983, ISBN 0-231-05392-4. 284p; illus, diags. Bibl (3), index (2).

A scholarly history of changing concepts of the universe, written by astronomers rather than historians, suitable for the general reader. Begins with prehistoric astronomy and cosmology, then moves forward to the present, examining the theories of the past in the light of current knowledge. The period up to Newton is covered in greater detail (eleven chapters) than subsequent developments (four chapters).

529 Ferris, Timothy
THE RED LIMIT: THE SEARCH FOR THE EDGE OF THE UNIVERSE (2nd ed.)
New York: Morrow Quill, 1983, ISBN 0-688-01836-X pbk. 286p; illus. Glos (3), bibl (2), index (2).

The story of the discovery of the expanding nature of the universe, including many quotations from the astronomers and other scientists who contributed to this discovery, or were involved in the redshift controversy, some based on personal interviews by the author. (1st ed. 1977 only seen).

530 Harrison, Edward R.
COSMOLOGY: THE SCIENCE OF THE UNIVERSE
Cambridge: Cambridge University Press, 1981, ISBN 0-521-22981-2. 430p; illus, diags. Bibl (x), index (3). Q

A thorough textbook for the non-science student and the interested lay reader, which provides a good introduction to past and present cosmological theory. Each of the 20 chapters is followed by 'reflections' intended to raise further questions. Includes good bibliographical references and a comprehensive index.

531 Heller, Michael, (A. Potocki, trans., George W. Collins, II, ed.)
ENCOUNTERING THE UNIVERSE
Tucson, AZ: Pachart, 1982, ISBN 0-912918-07-1 pbk. 110p. Index (1).

Translated from the Polish, and reproduced from typescript, this is a general, non-mathematical account of the development of cosmological thought.

532 Layzer, David
CONSTRUCTING THE UNIVERSE
New York: Scientific American, 1984, ISBN 0-7167-5003-1. 313p; illus

& diags (some col.). Index (2).

This attractively produced volume discusses the development of cosmological thought, focusing on the theories of space, time and gravitation of Newton and Einstein. Mathematical background knowledge is required for a full appreciation of this book.

533 Lovell, Bernard
EMERGING COSMOLOGY
New York: Columbia University Press, 1981, ISBN 0-231-05304-5; New York: Praeger, 1985, ISBN 0-03-001009-8 pbk. 208p.

A history of cosmology for the general reader, from Aristotle to current theories of the very early stages of the Big Bang.

534 Lovell, Bernard
IN THE CENTRE OF IMMENSITIES
New York: Harper & Row, 1978, ISBN 0-06-012716-3; London: Hutchinson, 1979, ISBN 0-09-136780-8; London: Granada, 1980, ISBN 0-586-08362-6 pbk. 174p. Index (2).

A general account of the development of cosmological thought, from the time of Copernicus to the Big Bang, which requires no special knowledge of physics or mathematics. The conclusion deals with 'the implications of [recent cosmological] discoveries for the development of life on Earth, and considers man as an element in the cosmic process of evolution'.

535 Munitz, Milton K.
SPACE, TIME AND CREATION: PHILOSOPHICAL ASPECTS OF SCIENTIFIC COSMOLOGY (2nd ed.)
New York: Dover, 1981, ISBN 0-486-24220-X. 184p. Bibl (x), index (1).

A non-mathematical look behind cosmological theories, which examines 'representative [theories], both historical and contemporary, and the methods which lie behind their construction'. This edition includes a new chapter, "Creationism and Big Bang cosmology".

Modern cosmology

536 Atkins, P.W.
THE CREATION
San Francisco: Freeman, 1981, ISBN 0-7167-1350-0. 132p. Bibl (2).

An unusual book about the universe, how it began, and why things are as they are (for example: why three spatial dimensions?). Based on the premise that the fundamental nature of the universe is far more simple to understand than we think. Rejects the idea that we need some sort of supreme being to explain the existence of the universe, instead explores the idea that the universe came into being because it could not fail to do so. The main text is on the right-hand pages, the left-hand side has notes for clarification.

537 Barrow, John D. and Joseph Silk
THE LEFT HAND OF CREATION: THE ORIGIN AND EVOLUTION OF THE EXPANDING UNIVERSE
New York: Basic Books, 1983, ISBN 0-456-03895-6; London: Heinemann, 1984, ISBN 0-434-04760-0; London: Allen & Unwin (Counterpoint), 1985, ISBN 0-04-523002-1 pbk. 256p; illus, diags. Glos (3), bibl (2), index (3).

A non-mathematical explanation of the ideas of modern cosmology, written in a readable, sometimes lyrical style. The subject matter is difficult to comprehend, but the authors come about as close as possible to making it comprehensible to the lay reader.

538 Boslough, John
BEYOND THE BLACK HOLE: STEPHEN HAWKING'S UNIVERSE (UK)/STEPHEN HAWKING'S UNIVERSE (US)
London: Collins, 1985, ISBN 0-00-217138-4; New York: Morrow, 1985, ISBN 0-668-03530-2. 127 + 8p; illus.

The life and, especially, the work of Professor Stephen Hawking, whose achievements in theoretical physics have been made despite the handicap of the wasting illness, motor neurone disease. This is a popular account of his cosmological theories and his work on the unification of the basic forces of nature. Includes an appendix by Hawking, "Is the end in sight for theoretical physics?: an inaugural lecture" (Cambridge, April 1980).

539 Davies, P.C.W. [Paul]
THE ACCIDENTAL UNIVERSE
Cambridge: Cambridge University Press, 1982, ISBN 0-521-24212-6; 0-521-28692-1 pbk. 139p; Bibl (2), index (2).

Presents the radical hypothesis that the physical structure of the universe is extremely contrived: physical systems owe their existence and form to combinations of improbable accidents. Concludes with a discussion of the only systematic attempt to explain this, the anthropic principle, which relates basic world features to our own existence as observers. A non-specialist treatment using only elementary algebra.

540 Davies, Paul
THE RUNAWAY UNIVERSE/STARDOOM: A SCIENTIFIC ACCOUNT OF THE BEGINNING AND END OF THE UNIVERSE (UK pbk)
London: Dent, 1978, ISBN 0-460-04286-6; London: Fontana, 1979, ISBN 0-00-635318-5 pbk; New York: Harper & Row, 1979, ISBN 0-06-010971-8; New York: Penguin, 1980, ISBN 0-14-005366-2 pbk. 205 + 8p; illus, diags. Index (2).

A popular account of modern theories of the history of the universe. Attempts 'to show how all physical systems, from a man to a galaxy, share the common, and somewhat enigmatic, quality of organization'.

541 Disney, Michael
THE HIDDEN UNIVERSE
London: Dent, 1984, ISBN 0-460-04441-9; New York: Macmillan, 1985, ISBN 0-02-531670-2. 216 + 8p; illus, diags. Bibl (2), index (1).

A rather demanding (for the general reader) discussion of a key cosmological question: will the universe continue to expand forever or will gravitational forces pull all matter back together? The answer lies in the mass of the universe: the density of matter detected by astrophysicists is far less than is required if the expansion is to be halted; the rest of the mass is 'hidden', but is it sufficient?

*542 Feibleman, James K.
THE NEW COSMOLOGY
St. Louis, MO: Green, 1982, ISBN 0-87527-171-5. 224p.

543 Gingerich, Owen, (introduced by)
COSMOLOGY + 1
San Francisco: Freeman, 1977, ISBN 0-7167-0043-3; 0-7167-0042-5 pbk. 113p; illus & diags (some col.). Bibl (1), index (2). Q

A collection of articles reprinted from *Scientific American*; eight are devoted to cosmology, two to black holes, and one to the search for extraterrestrial intelligence. Two of the pieces are rather dated.

544 Gribbin, John (ed.)
COSMOLOGY TODAY: A *NEW SCIENTIST* GUIDE
London: IPC Magazines, 1982, ISBN 0-85037-518-5 pbk. 64p; illus, diags. Q

A collection of 18 articles from *New Scientist*, divided into five parts:

1. Einstein's universe, 2. the modern universe (its age, beginning, galaxy formation, and end), 3. violence in the universe (quasars, black holes, galactic nuclei), 4. the anthropic principle (philosophic essays on man's place in the universe), and 5. the devil's advocates (dissenting views). Unfortunately the book lacks an index.

545 Gribbin, John
WHITE HOLES: COSMIC GUSHERS IN THE UNIVERSE
St. Albans: Granada (Paladin), 1977, ISBN 0-586-08263-8 pbk; New York: Delacorte (Eleanor Friede), 1977, ISBN 0-440-09529-8. 200 + 8p; illus, diags. Glos (2), bibl (2), index (2).

A popular account of modern cosmological thought, which concentrates on speculative ideas, some drawn from science fiction. A white hole is the opposite of a black hole, it pours matter into the universe rather than sucking it out. An appendix asks, "Is our Sun a normal star?"

546 Hoyle, Fred
FACTS AND DOGMAS IN COSMOLOGY AND ELSEWHERE: THE REDE LECTURE 1982
Cambridge: Cambridge University Press, 1982, ISBN 0-521-27268-8 pbk. 16p.

This transcript of a lecture given to a general audience presents a brief explanation of cosmology, discusses various theories, and defends the validity of the steady state theory. The author touches on biology, evolution and religion.

547 Huff, Douglas and Omer Prewitt (ed.)
THE NATURE OF THE PHYSICAL UNIVERSE
New York: Wiley, 1979, ISBN 0-471-03190-9. 140p.

Papers presented at the 1976 Nobel Conference in Minnesota. The contributors include Murray Gell-Man, Fred Hoyle and Steven Weinberg. Technical in parts, no index.

548 Islam, Jamal N.
THE ULTIMATE FATE OF THE UNIVERSE
Cambridge: Cambridge University Press, 1983, ISBN 0-521-24814-0. 155p; illus, diags. Glos (2), bibl (1), index (2).

A descriptive account for the general reader. Technical terms are described as simply as possible, while attempting to avoid oversimplification. Covers the present state and structure of the universe, before speculating on its final evolutionary stages. Relies mainly on the Big Bang model, but also discusses steady-state cosmology.

549 Johnston, Alan
THE COSMIC ECOSYSTEM: A REVISED VIEW OF COSMOLOGY
London: Wildwood House, 1980, ISBN 0-7045-3044-9. 194p; diags. Bibl (1), index (2).

A dissenting view of the history of the universe, written in non-technical language. Following three chapters of background information, the author presents his argument in favour of the steady state theory; he argues that quasars are not, as generally believed, very distant objects, and proposes the theory of 'quantum repulsive gravitation' to explain how the universe expands.

*550 Morris, Richard
THE FATE OF THE UNIVERSE
New York: Playboy Press, 1982, ISBN 0-87223-748-6. 174p.

551 Novikov, I.D., (M.M. Basko, trans.)
EVOLUTION OF THE UNIVERSE
Cambridge: Cambridge University Press, 1983, ISBN 0-521-24129-4. 176p; diags. Bibl (x).

This outline of modern cosmology assumes 'no special knowledge beyond school mathematics and physics'. It is written from a Soviet point of view, presenting the results of Russian work to a wider readership. This is a translation of the first Soviet edition, 1979; the English edition has been updated to Spring 1981.

552 Parish, Leonard
THE THEORY OF COSMIC ABERRATION: A NEW INTERPRETATION OF THE HUBBLE RED-SHIFTS
Luton: Cortney, 1981, ISBN 0-904378-11-X. 44p; diags.

The author suggests that the observed redshifts of extra-galactic objects are not due entirely to recessional velocities, but can be partly accounted for by the movement of observers due to the rotational speed of the galaxy. Uses some algebra.

553 Sciama, D.W.
MODERN COSMOLOGY
Cambridge: Cambridge University Press, 1982, ISBN 0-521-28721-9 pbk (reprint of 1975 printing, with updating note, of corrected printing 1972 of 1st ed. 1971). 214 + 16p; illus, diags. Bibl (1), indexes (2).

A general survey of cosmology for readers with background knowledge of mathematics and physics. The first half of the book discusses the major components of the universe: stars, the Milky Way, other galaxies, radio galaxies and quasars. The second half deals with models of the universe, the search for intergalactic atomic and ionized hydrogen, the Big Bang and helium formation, the cosmic 3°K microwave radiation and isotropy.

554 Silk, Joseph
THE BIG BANG: THE CREATION AND EVOLUTION OF THE UNIVERSE
San Francisco: Freeman, 1980, ISBN 0-7167-1084-6; 0-7167-1085-4 pbk. 394p; illus, diags. Bibl (1), glos (3), index (2).

This introductory work is fairly technical in places; it is probably intended as an introduction for students of astronomy and physics, but possibly also for the serious amateur or general reader. Provides an introduction to cosmology, and discusses evidence for the Big Bang, cosmological models and the evolution of the universe. Includes 34 pages of mathematical notes and an 18-page glossary.

555 Toben, Bob
THE HOME OF THE STARS: THE ILLUSTRATED STORY OF THE UNIVERSE AND ITS BEGINNINGS
New York: Dutton, 1982, ISBN 0-525-47629-6 pbk. 190p; illus. Q

A sketchily-drawn, comic-book-style account of cosmological ideas.

556 Trefil, James S.
THE MOMENT OF CREATION: BIG BANG PHYSICS FROM BEFORE THE FIRST MILLISECOND TO THE PRESENT UNIVERSE
New York: Scribner's, 1983, ISBN 0-684-17963-6; New York: Macmillan, 1984, ISBN 0-02-096770-5 pbk. 234p; diags. Glos (2), index (2).

A largely non-mathematical account of current cosmological thought. The first part covers briefly the history of the discovery of events ever closer to the instant of the Big Bang. The second part explains the new physics of gauge theories, which may prove too difficult for the general reader, but the author provides chapter summaries for those wishing to move on quickly to the final part, which, using an understanding of Grand Unified Theories (of the basic forces of nature), attempts to look beyond the first millisecond after the Big Bang to approach the moment of creation itself. Clearly and systematically presented.

557 Weinberg, Steven
THE FIRST THREE MINUTES: A MODERN VIEW OF THE ORIGIN OF THE UNIVERSE
New York: Basic Books, 1976, ISBN 0-465-02435-1; London: André Deutsch, 1977, ISBN 0-233-96906-3; London: Fontana (Flamingo), 1983, ISBN 0-00-654024-4 pbk. 188 + 10p; illus, diags. Glos (2), bibl (2), index (2).

A non-technical account of recent advances in our understanding of events shortly after the Big Bang. Not an overview of the whole field of cosmology, but does include some background information. There is an 11-page mathematical supplement. The 1983 paperback edition includes a new afterword.

*558 Willcox, P.J.
MODERN COSMOLOGY
Huntington, IN: P.J. Willcox, 1981, pbk. 96p.

Four lectures on the historical evolution of cosmological thought, the Big Bang, stellar evolution, and the future of cosmology, by an aerospace researcher. (From *S&T* Dec. 1982, p 560).

The Universe

559 Asimov, Isaac
THE COLLAPSING UNIVERSE
New York: Walker, 1977, ISBN 0-8027-0486-7; London: Hutchinson, 1977, ISBN 0-09-131770-3; London: Transworld (Corgi), 1978, ISBN 0-552-10884-7 pbk; New York: Pocket Books, 1981, ISBN 0-671-44868-4 pbk. 204p. Index (2).

A general account of our knowledge of the structure of the universe, set in a cosmological framework, presented in an entertaining style. Less demanding than Motz (564).

*560 Fritzsch, Harald
THE CREATION OF MATTER: THE UNIVERSE FROM BEGINNING TO END
New York: Basic Books, 1984, ISBN 0-465-01446-1. 272p.

561 Gibilisco, Stan
BLACK HOLES, QUASARS AND OTHER MYSTERIES OF THE UNIVERSE
Blue Ridge Summit, PA: TAB Books, 1984, ISBN 0-8306-1525-3 pbk. 197 + 8p; illus (some col.), diags. Bibl (1), index (2).

A general introduction to the universe and cosmic phenomena. Includes brief coverage of the Solar System, the possibility of extraterrestrial life and interstellar travel.

562 Heidmann, Jean, (Maureen Schaeffer and Ann Boesgaard, trans.)
EXTRAGALACTIC ADVENTURE: OUR STRANGE UNIVERSE
Cambridge: Cambridge University Press, 1982, ISBN 0-521-23571-5; 0-521-28045-1 pbk (trans. of 1st French ed. 1979). 174p; illus, diags.

This brief beginner's guide to the universe is written in a breathless style, not quite idiomatically translated. Provides a painless introduction to technical terms and concepts such as quantum mechanics and relativity. A book to stir the imagination of any newcomer, but lacks a guide to further reading, as well as an index.

563 Henbest, Nigel and Heather Couper
THE RESTLESS UNIVERSE
London: George Philip, 1982, ISBN 0-540-01069-3. 214 + 16p; col. illus, diags. Index (2).

This general survey of our knowledge of the universe concentrates on recent discoveries. About a third of the book is devoted to the Solar System, the rest to the stars, galaxies and cosmology. Written in a readable style, the book contains some very good, though rather small colour illustrations, but has few explanatory diagrams.

564 Motz, Lloyd
THE UNIVERSE
New York: Scribner's, 1975, ISBN 0-684-14239-2; London: Millington, 1976, ISBN 0-86000-071-0; London: Sphere (Abacus), 1977, ISBN 0-349-12405-1 pbk. 342 + 16p; illus. Bibl (1), index (3).

A general account of our knowledge of the universe. Less demanding than Narlikar (565), and with a different emphasis: this book describes the evolution of the universe from its beginning as far as the origin of life.

565 Narlikar, Jayant
THE STRUCTURE OF THE UNIVERSE
London: Oxford University Press, 1978, ISBN 0-19-217653-6; 0-19-289082-4 pbk (corrected reprint of 1st ed. 1977). 264p; illus, diags. Glos (2), bibl (1), index (2).

An introductory work for undergraduates in physics or astronomy, and any interested reader with a knowledge of elementary algebra. Covers stellar evolution, our galaxy, other galaxies, quasars, cosmology, gravitation, and 'the universe and the arrow of time'.

566 Nicolson, Iain and Patrick Moore
THE UNIVERSE
London: Collins, 1985, ISBN 0-00-217397-2. 256p; illus & diags (mostly col.). Bibl (1), glos (1), index (2). Q

A colourful, attractively presented survey of our knowledge of the universe for the general reader and the beginning amateur. The book starts with the Solar System and moves ever outward, but pauses to bring general principles together.

567 Shipman, Harry L.
BLACK HOLES, QUASARS AND THE UNIVERSE (2nd ed.)
Boston: Houghton Mifflin, 1980, ISBN 0-395-29916-0; 0-395-28499-6 pbk. 320p; illus, diags. Glos (2), bibl (2), index (2).

An introductory work for the lay reader, student and non-specialist scientist, which highlights the changes that took place in astronomy in the 1960s and 1970s, and shows the renewed importance of cosmology.

Galaxies and Quasars

568 Bok, Bart J. and Priscilla F.
THE MILKY WAY (5th ed.)
Cambridge, MA and London: Harvard University Press, 1981, ISBN 0-674-57503-2. 356p; illus, diags. Index (3).

The latest edition of a classic introductory work, first published in 1941. Begins with a discussion of the observational appearance of the Milky Way and methods of observation, then considers what is measured and how the data are interpreted. Further chapters cover types of stars and clusters, motion within the galaxy and of the galaxy itself, the galactic nucleus, interstellar gas, nebulae, the spiral structure, and change in the galaxy.

569 Ferris, Timothy
GALAXIES
San Francisco: Sierra Club Books, 1980, ISBN 0-87156-273-1; London: Thames & Hudson, 1980, ISBN 0-500-01248-2; New York: Stewart, Tabori & Chang, 1982, ISBN 0-941434-01-X; 0-941434-02-8 pbk. 182p; illus & diags (mostly col.). Bibl (1), glos (2), index (2). Q (1980 ed. 34 × 37 cm).

A very large, beautifully illustrated coffee-table book, which takes the reader on a tour of the universe from the Milky Way outward.

The text, while overshadowed by the magnificent, often full-page colour illustrations, is not insubstantial: the author provides a clear explanation of the important concepts aided by a number of very good diagrams. The book was updated for the smaller-sized 1982 edition.

570 Hodge, Paul W. (compiler)
THE UNIVERSE OF GALAXIES
New York: Freeman, 1984, ISBN 0-7167-1676-3 pbk. 113p; illus, diags. Bibl (2), index (2). Q

A collection of nine articles reprinted from *Scientific American*, 1973–83, showing the development of our knowledge of galaxies. Coverage ranges from normal galaxies, such as the Milky Way and the Andromeda galaxy, to superclusters, cosmic jets and quasars.

571 Hoyle, Fred
THE QUASAR CONTROVERSY RESOLVED
Cardiff: University College Cardiff Press, 1981, ISBN 0-906449-28-6 pbk. 80p; illus, diags. Index (1).

The author presents a possible solution which is independent of any cosmological model, concluding that quasars are not such distant objects as is generally believed. Uses algebra, and includes autobiographical reminiscence. Ends with a technical appendix by Hoyle and J.V. Narlikar.

572 Kaufmann, William J., III
GALAXIES AND QUASARS
San Francisco: Freeman, 1979, ISBN 0-7167-1133-8; 0-7167-1134-6 pbk. 226 + 4p; illus (some col.), diags. Bibl (1), index (2).

A beginner's guide to galactic astronomy, with some coverage of cosmology. Profusely illustrated and clearly explained.

573 Kühn, Ludwig
THE MILKY WAY: THE STRUCTURE AND DEVELOPMENT OF OUR STAR SYSTEM
Chichester: Wiley, 1982, ISBN 0-471-10277-6. 151p; illus, diags. Index (1).

This uncredited translation of the German edition of 1978 presents a survey of our knowledge of the Galaxy for the general reader. It was very poorly received by reviewers.

574 Mitton, Simon
EXPLORING THE GALAXIES
London: Faber, 1976, ISBN 0-571-10925-X; New York: Scribner's,

1977, ISBN 0-684-14862-5; 0-684-15579-6 pbk. 206 + 16p; illus, diags. Bibl (1), indexes (2).

A non-technical introduction to galactic astronomy.

575 Rigutti, Mario, (Mirella Giacconi, trans.)
A HUNDRED BILLION STARS
Cambridge, MA: MIT Press, 1984, ISBN 0-262-18111-8 (trans. of 1st Italian ed. 1978). 284 + 6p; illus (some col.), diags. Index (3).

A general account of our knowledge of the contents of the Galaxy. The book's three parts cover the Solar System, the Galaxy, and stellar evolution.

576 Tayler, R.J.
GALAXIES: STRUCTURE AND EVOLUTION
London: Wykeham, 1978, ISBN 0-85109-680-8; 0-85109-670-0 pbk. 203p; illus, diags. Bibl (1), index (1).

An account of our knowledge of the galaxies, intended to bridge the gap between senior secondary school and undergraduate levels. Requires a knowledge of algebra. Attempts to provide a precise treatment of the subject, but owing to the uncertainty of available data few numerical values are quoted.

Stars, Stellar Evolution and Black Holes

577 Clark, David
SUPERSTARS
London: Dent, 1979, ISBN 0-460-04384-6; New York: McGraw-Hill, 1984, ISBN 0-07-011529-X. 175 + 8p; illus, diags. Index (2).

A popular account of supernovae: how they occur, their influence on history, and their study. The book has a strong historical flavour, drawing on research in many fields, including archaeology, astronomy, physics, climatology and paleobiology. Discusses the effects of supernova explosions, particularly with regard to stellar and planetary formation, and considers the possible effects on the Solar System of a nearby supernova. (*See also* 464).

*578 Cooke, Donald A.
THE LIFE AND DEATH OF STARS
New York: Crown, 1985, ISBN 0-517-55268-X.

579 Gallant, Roy A.
FIRES IN THE SKY: THE BIRTH AND DEATH OF STARS
New York: Four Winds Press, 1978, ISBN 0-590-07475-X. 162p; illus, diags. Glos (3), index (2).

This basic introduction is aimed primarily at the younger reader, but may appeal to the adult novice. The author focuses on the Sun in discussing stellar evolution and the physical properties of stars, but also considers, in separate sections, the different types of star.

580 Greenstein, George
FROZEN STAR
New York: Freundlich Books, 1983, ISBN 0-88191-011-2; London: Macdonald, 1984, ISBN 0-356-10588-1; New York: New American Library (Plume), 1985, ISBN 0-452-25693-3 pbk. 274 + 8p; illus, diags. Index (2).

An account of pulsars, black holes and neutron stars for the general reader. Beginning with the discovery of pulsars, the author highlights the process of discovery and its excitement in this and subsequent advances in our knowledge of the components of the Galaxy.

581 Kaufmann, William, J., III
BLACK HOLES AND WARPED SPACETIME
San Francisco: Freeman, 1979, ISBN 0-7167-1152-4; 0-7167-1153-2 pbk. 221 + 16p; illus (some col.), diags. Index (2).

A fairly brief, non-technical, profusely illustrated account of a relatively new area of astrophysics which impinges on the field of cosmology. Begins with a description of the process of stellar evolution, and ends with speculation on the beginning and end of the universe.

582 Kaufmann, William J., III
STARS AND NEBULAS
San Francisco: Freeman, 1978, ISBN 0-7167-0081-6; 0-7167-0085-9 pbk. 204 + 8p; illus (some col.), diags. Bibl (1), index (2).

An introduction for beginners. Following a discussion of stars in general the author deals with the laws of light and the significance of starlight, then covers the Sun, stellar evolution, pulsars, neutron stars and black holes. Includes basic monthly star maps.

583 Kippenhahn, Rudolf, (Jean Steinberg, trans.)
100 BILLION SUNS: THE BIRTH, LIFE, AND DEATH OF THE STARS
New York: Basic Books, 1983, ISBN 0-465-05263-0; London:

Weidenfeld & Nicolson, 1983, ISBN 0-297-78248-7; London: Allen & Unwin (Counterpoint), 1985, ISBN 0-04-523003-X pbk. 264 + 6p; illus (some col.), diags. Index (1).

A good general account for the lay reader by the director of the Max Planck Institute for Astrophysics. Covers stellar evolution, types of star, and the end products of evolution: supernovae, pulsars and black holes. This translation of the German edition of 1980 incorporates more recent knowledge.

584 Meadows, A.J.
STELLAR EVOLUTION (2nd ed.)
Oxford: Pergamon Press, 1978, ISBN 0-08-021668-4; 0-08-021669-2 pbk. 171p; diags. Bibl (1), index (1).

A non-mathematical account, fairly demanding, but less so than Tayler (591).

585 Mitton, Simon
THE CRAB NEBULA
London: Faber, 1979, ISBN 0-571-10367-7; New York: Scribner's, 1979, ISBN 0-684-16077-3; 0-684-17531-2 pbk. 194p; illus, diags. Bibl (1), glos (2), index (1).

Aims 'to give a biography of the Crab Nebula' based on the 'several hundred research papers and monographs devoted to it in particular and supernovae in general'. A non-mathematical account which deals with the influence of the Crab Nebula on the history of modern astronomy, the structure of the Nebula, and the physics of the pulsar.

586 Moore, Patrick and Iain Nicolson
BLACK HOLES IN SPACE
London: Orbach & Chambers, 1979, ISBN 0-427-00449-7 pbk (reprint of 1st ed. 1974). 126p; diags. Index (1).

A general account for the lay reader, now somewhat dated.

587 Murdin, Paul and Lesley
THE NEW ASTRONOMY
London: Reference International, 1978, ISBN 0-905154-06-1; New York: Crowell, 1978, ISBN 0-690-01474-0. 215p; illus. Bibl (1), index (1).

A non-technical account of research into stellar explosions, particularly supernovae (the title is too broad), as examples of the progress made possible by the use of radio astronomy. The authors focus on the supernova responsible for the Crab Nebula, and on the Nebula itself. They also discuss other aspects of supernova research, for example, how it sheds light on the question of the formation of the elements and the origin of cosmic rays.

Not yet published at the time of writing is the authors' *Supernovae* (Cambridge: Cambridge University Press, 1985, ISBN 0-521-30038-X), which may be a revised version of this book.

588 Payne-Gaposchkin, Cecilia
STARS AND CLUSTERS
Cambridge, MA: Harvard University Press, 1979, ISBN 0-674-83440-2. 262p; illus, diags. Indexes (3).

A descriptive account of the stars of our Galaxy: their formation and evolution, the different types, groups of stars, how they are studied and what can be discovered about them. Likely to prove too demanding for the complete beginner. Includes a 21-page AAVSO star atlas (epoch 1900.0).

589 Spitzer, Lyman, jr.
SEARCHING BETWEEN THE STARS
New Haven and London: Yale University Press, 1982, ISBN 0-300-02709-5; 1984, ISBN 0-300-03247-1 pbk. 179p; illus, diags. Bibl (2), index (2).

A descriptive account of our current knowledge of interstellar matter, based on a series of lectures given in 1978. The emphasis is on discoveries made through new observational techniques, particularly the results of the Copernicus satellite (OAO 3) launched in 1972, which studied interstellar gas. The use of mathematics is kept to a minimum, and summaries are provided at the end of each chapter.

590 Sullivan, Walter
BLACK HOLES: THE EDGE OF SPACE, THE END OF TIME
Garden City, NY: Doubleday (Anchor Press), 1979, ISBN 0-385-07516-2; New York: Warner, 1980, ISBN 0-446-32281-0 pbk. 304p; illus, diags. Bibl (3), index (2).

A book for the general reader by a journalist. Much of the text is devoted to historical background material, dwelling on the personalities of the astronomers involved. Begins with Einstein's predictions concerning the curved nature of space, and continues with the discovery of the scale of the universe and its expanding nature. Covers also peripheral topics such as quasars, x-ray astronomy and cosmology, before turning to black holes.

591 Tayler, R.J.
THE STARS: THEIR STRUCTURE AND EVOLUTION (rev. ed.)
London: Wykeham, 1978, ISBN 0-85109-110-5 pbk. 208p; diags. Bibl (1), index (1).

Part of a series intended 'to introduce the present state of science as a university subject to students approaching or starting their university careers'. Uses mathematics appropriate to this level.

592 Taylor, John
BLACK HOLES: THE END OF THE UNIVERSE?
Glasgow: Fontana, 1980, ISBN 0-00-613509-9 pbk (reprint of 1st ed. 1974). 190p. Bibl (1), indexes (2).

This early, highly speculative, popular account was much criticized by reviewers.

6 The Solar System

General

593 Beatty, J. Kelly, Brian O'Leary and Andrew Chaikin (ed.)
THE NEW SOLAR SYSTEM (2nd ed.)
Cambridge: Cambridge University Press, 1982, ISBN 0-521-24988-0; 0-521-27114-2 pbk; Cambridge, MA: Sky, 1982, ISBN 0-933346-36-0; 0-933346-37-9 pbk. 240p; illus & diags (mostly col.). Bibl (2), index (2). Q

Published only a year after the first edition, this survey of our knowledge of the Solar System has been extensively revised throughout. The 20 chapters by specialists cover a wide range of topics, the method of arrangement of the book reflecting the increasing importance of comparative planetology, the study of features, processes and phenomena common to more than one body, rather than the study of each body in isolation. So, though there are chapters devoted to the major planets, there are also chapters covering, for example, magnetospheres and the interplanetary medium, surfaces of the terrestrial planets and planetary rings. In addition to the planets there is good coverage of the Sun, planetary satellites, comets and meteorites. This edition includes a 14-page section of maps in addition to the many excellent illustrations in the text. The book is written at a level which would not suit the beginner.

594 Bova, Ben and Trudy E. Bell (ed.)
CLOSEUP: NEW WORLDS
New York: St. Martins, 1977, ISBN 0-312-14490-3; 0-312-14491-1 pbk. 222 + 4p; illus (some col.), diags. Glos (2). Q

A collection of seven articles, all but one of which are new to this book, describing recent developments in the study of the Solar System. Some are by science fiction writers: the intention of the editors being to increase the reader's interest by including with each article speculation on future exploration and the human uses of the 'new worlds'.

595 Chapman, Clark R.
PLANETS OF ROCK AND ICE: FROM MERCURY TO THE MOONS OF SATURN
New York: Scribner's, 1982, ISBN 0-684-17484-7. 222 + 32p; illus. Index (3).

A largely non-technical survey of planetary science, which conveys clearly the importance of such research and makes the reader aware of the uncertainty surrounding newly gained knowledge. This is a revised version of *The inner planets* (Scribner's, 1977), updated to incorporate the results of the Voyager missions to Jupiter and Saturn.

596 Firsoff, V.A.
THE SOLAR PLANETS
Newton Abbot: David & Charles, 1977, ISBN 0-7153-7352-8; New York: Crane, Russak, 1977, ISBN 0-8448-0964-0. 184 + 6p; illus, diags. Bibl (3), indexes (2).

A non-technical account of our knowledge of the Solar System, published shortly after the Viking mission, so not surprisingly there is a long chapter on Mars.

597 Francis, Peter
THE PLANETS: A DECADE OF DISCOVERY
Harmondsworth: Penguin, 1981, ISBN 0-14-022053-4 pbk. 411p; illus, diags. Bibl (2), index (2).

A non-technical account of our knowledge of the planets (excluding Earth), by a geologist. Provides historical background and describes the methods used to obtain information about the planets, from Earth and from spacecraft. Good coverage of physical features.

598 Hancock, C.B.
OUTREACH FOR THE PLANETS
Ilfracombe: Arthur H. Stockwell, 1979, ISBN 0-950668-40-0 pbk. 147p; diags. Index (1).

Based on a series of lectures given by 'a man knowledgeable in several sciences', this book is intended for the reader with no initial knowledge of astronomy. The author points out that he takes great pains to achieve accuracy, and stresses the importance of space exploration.

599 Jackson, Joseph H. and John H. Baumert
PICTORIAL GUIDE TO THE PLANETS (3rd ed.)
New York: Harper & Row, 1981, ISBN 0-06-014869-1. 246 + 4p; illus

(some col.), diags. Index (2). Q

A book for the general reader which incorporates much recent knowledge, but was published before the Voyager results from Jupiter and Saturn were available. Includes chapters on life (on Earth and elsewhere), artificial satellites and space probes, and 'man in space', in addition to the coverage of planetary science.

600 Moore, Patrick, Garry Hunt, Iain Nicolson and Peter Cattermole
THE ATLAS OF THE SOLAR SYSTEM
London: Mitchell Beazley, 1983, ISBN 0-85533-468-1. New York: Rand McNally, 1983, ISBN 0-528-81122-3. 464p; illus & diags (some col.). Glos (1), bibl (2), index (3). Q

Published under the auspices of the Royal Astronomical Society, this book incorporates material published earlier in a series of smaller atlases intended for both professional and amateur astronomers, and some new material. The previous volumes were *The Sun* (633), *The Moon* (676), *Jupiter* (699), and *Saturn* (700). Each body is treated separately, the format being the same as that used earlier: a series of two-page spreads covering selected topics. Ends with a chapter on space exploration. The standard two-page glossary of the earlier volumes has been replaced by an abridged single-page version.

601 Murray, Bruce, (selected and introduced by)
THE PLANETS: READINGS FROM *SCIENTIFIC AMERICAN*
San Francisco: Freeman, 1983, ISBN 0-7167-1467-1; 0-7167-1468-X pbk. 132p; illus & diags. (some col.). Bibl (2), index (1). Q

A collection of ten articles, 1975–82, including one on Mercury from *The Solar System* (Freeman, 1975), which was a reprint of the special September 1975 issue of *Scientific American*. Covers the atmospheres and surfaces of Venus and Mars, Jupiter and Saturn, the Galilean moons of Jupiter, Titan, the moons of Saturn, and rings in the Solar System, ending with a note on future planetary exploration.

A popular-level periodical omitted from the main listing is *The Planetary Report* (The Planetary Society, PO Box 91687, Pasadena, CA 91109, USA), 1981–, bi-monthly, ISSN 0736-3680. About 20p; illus & diags (mostly col.), Q.

602 Smoluchowski, Roman
THE SOLAR SYSTEM
New York: Scientific American, 1983, ISBN 0-7167-1492-2; 0-7167-1493-0 pbk. 174p; illus (some col.), diags. Bibl (1), index (2).

An introduction for the general reader, below the level of a textbook, but not over-simplistic. The text is arranged in the form of three essays: the birth of the Sun, planetary and satellite formation, the fate of the Solar System; and five chapters covering the Sun, planets and other bodies. An afterword discusses life in the Solar System.

Late entry: Kendrick Frazier, *Solar System* (Alexandria, VA: Time-Life, 1985, ISBN 0-8094-4530-1; Amsterdam: Time-Life, 1985, ISBN 0-7054-0755-1, 176p), a colourful, popular work.

603 von Braun, Wernher and Frederick I. Ordway, III, (editorial assistance by Eric Burgess)
NEW WORLDS: DISCOVERIES FROM OUR SOLAR SYSTEM
Garden City, NY: Doubleday (Anchor Press), 1979, ISBN 0-385-14065-7. 284p; illus (some col.), diags. Index (3). Q

A non-technical account of the knowledge of the Solar System gained in the space age, but written before the Voyager spacecraft reached Jupiter and Saturn.

604 Whipple, Fred L.
ORBITING THE SUN: PLANETS AND SATELLITES OF THE SOLAR SYSTEM
Cambridge, MA: Harvard University Press, 1981, ISBN 0-674-64125-6. 338 + 4p; illus (some col.), diags. Index (2).

This introductory work for the general reader is a little more rigorous than most such books; it provides an introduction to positional astronomy as well as describing the features and physical processes of the planets and moons. There is a chapter on observing the Moon, and the appendices include the ‘planet finder’, a series of tables used in conjuction with the star chart provided to find the positions of the planets, 1981–92. This is a revised version of *Earth, Moon and planets* (3rd ed. 1968; 1st ed. 1941).

Textbooks

For a textbook which deals only with the terrestrial planets *see* (647).

605 Cole, G.H.A.
THE STRUCTURE OF PLANETS
London: Wykeham, 1978, ISBN 0-85109-610-7; 0-85109-600-X. 232p; diags. Bibl (1), index (2).

An introduction to planetary science intended to bridge the gap between senior secondary school and undergraduate levels. Coverage of surface geologies and atmospheres is restricted, and there is an introductory account only of some aspects of the study of planetary interiors. The first part of the book deals with general principles: gravitation and planetary orbits, and a description of the appearance of each body. The second part looks at the planets in more detail and considers the evidence obtained from exploration by spacecraft.

606 Hartmann, William K.
MOONS AND PLANETS (2nd ed.)
Belmont, CA: Wadsworth, 1983, ISBN 0-534-00719-8. 509p; illus, diags. Bibl (3), index (2).

This textbook of planetary science is intended for the general reader as well as the non-science undergraduate, and may be used with or without the mathematics included. Arrangement is by topic rather than by taking each body individually: following an overview of the subject the book covers celestial mechanics, the formation of the stars and planetary material, the formation of planets and satellites, meteorites and meteoritics, asteroids, comets, planetary interiors, planetary surfaces, atmospheres, and life. Includes chapter summaries, lists of key terms and problems.

607 Jones, Barrie William
THE SOLAR SYSTEM
Oxford: Pergamon Press, 1984, ISBN 0-08-026496-4; 0-08-026495-6 pbk. 336p; illus, diags. Bibl (2), index (2).

An introductory text which should be accessible to readers with little background in science and mathematics; most of the necessary background information is presented in specially marked sections. Thorough and well presented, the book is arranged by planet (including Earth) or planetary system, with separate chapters on topics such as impact cratering, planetary rings, and the origin and end of planetary systems. Includes questions.

608 Kaufmann, William J., III
EXPLORATION OF THE SOLAR SYSTEM
New York: Macmillan, 1978, ISBN 0-02-362140-0. 575 + 12p; illus (some col.), diags. Glos (3), index (3).

This text is designed to 'meet the need for a readable comprehensive book on the solar system sciences. It is written primarily for the liberal arts major at typical colleges and universities'. It is also suitable as an introduction for the interested lay reader. Covers history (of observations, and the development of physical principles), the changing sky, light and matter, the Sun, Earth, Moon, Mars, Venus, Mercury, Jupiter, the outer planets, asteroids, comets and meteorites, and the evolution of the Solar System.

609 Wood, John A.
THE SOLAR SYSTEM
Englewood Cliffs, NJ: Prentice-Hall, 1979, ISBN 0-13-822015-8 pbk. 196p; illus, diags. Glos (1), bibl (1), index (2).

An introduction for students below the undergraduate level which uses elementary algebra. Following a basic account of planetary motion, the author concentrates on a discussion of the composition of the planets, before turning to the origin of the Solar System.

Pictorial works

The books listed here emphasize illustrative matter, but this does not necessarily imply that their text is insignificant.

610 [Beet, E.A. and H.G. Miles, (compilers)]
THE SOLAR SYSTEM
London: British Astronomical Association, 1980, pbk. 20 leaves; illus (some col.).

A collection of 21 sets of astronomical photographs intended for educational purposes. All but two are printed on one side of each page, with a brief descriptive caption below, to allow cutting out and filing.

611 Briggs, Geoffrey and Fredric Taylor
THE CAMBRIDGE PHOTOGRAPHIC ATLAS OF THE PLANETS
Cambridge: Cambridge University Press, 1982, ISBN 0-521-23976-1. 225p; illus (mostly col.). Q

A collection of photographs and large-scale maps of the planets as far as Saturn, incorporating close-up views of the Jovian and Saturnian systems from the Voyager flybys. The accompanying text and captions attempt to summarize the current state of our knowledge of the planets. An index and a guide to further reading would have been useful additions to this book.

612 Doherty, Paul
ATLAS OF THE PLANETS
London: Hamlyn, 1980, ISBN 0-600-30439-6; New York: McGraw-Hill, 1980, ISBN 0-07-017341-9. 143p; illus (mostly col.). Glos (1), index (1). Q

An introductory work which provides much background information for amateur astronomers interested in the study of the planets. Covers the history of observation and discovery, and gives hints for present-day observers. Includes many drawings based on the author's own telescopic work, and others from imagination, but based on available data.

613 Hardy, David A.
ATLAS OF THE SOLAR SYSTEM
Tadworth: World's Work, 1981, ISBN 0-437-06540-5. 96p; illus & diags. (mostly col.). Glos (1), index (2). Q

An introductory account which provides a showcase for the author's astronomical artwork. Arrangement is by object. The text, which is quite substantial, concentrates on recent findings.

*614 Hartmann, William K., Ron Miller and Pamela Lee
OUT OF THE CRADLE: EXPLORING THE FRONTIERS BEYOND EARTH
New York: Workman, 1984, ISBN 0-89480-770-6 pbk. 190p; col. illus. Bibl, index.

This is not merely a popular introduction to the Solar System but a catalogue of possible planetary missions of the next century or two, visualized through space art. The paintings are by Hartmann, Miller and Lee; the text by astronomer Hartmann discusses the relationship of space exploration to the problems of humans on Earth. (From *S&T* April 1985, p 320–21).

615 Kaufmann, William J., III
PLANETS AND MOONS
San Francisco: Freeman, 1979, ISBN 0-7167-1042-2; 0-7167-1040-4 pbk. 219 + 4p; illus (some col.), diags. Bibl (2), index (2).

A brief, non-technical, profusely illustrated introduction for the beginner.

616 Miller, Ron and William K. Hartmann
THE GRAND TOUR: A TRAVELER'S GUIDE TO THE SOLAR SYSTEM (US)/THE TRAVELLER'S GUIDE TO THE SOLAR SYSTEM (UK)
New York: Workman, 1981, ISBN 0-89480-147-3; 0-89480-146-5 pbk;

London: Macmillan, 1981, ISBN 0-333-32694-6; 0-333-32695-4 pbk. 192p; illus (mostly col.), diags. Glos (1), bibl (1), index (2).

This well produced introductory work features art mostly by Miller, but also includes some samples of the work of astronomer and artist Hartmann, as well as a few photographs. The text provides a concise account of the current state of our knowledge of the Solar System.

617 Pizzey, Stephen
EXPLORING: THE SOLAR SYSTEM
London: HMSO, 1977, ISBN 0-11-290277-4 pbk. 28p; illus (some col.), diags. Bibl (1).

Produced to accompany the Exploration exhibition which opened at the Science Museum in London in 1977, this booklet provides a concise introduction for beginners of all ages.

618 Ryan, Peter and Ludek Pesek
SOLAR SYSTEM
London: Allen Lane, 1978, ISBN 0-7139-1043-7; New York: Viking, 1979, ISBN 0-670-65636-4. 224p; illus & diags (mostly col.). Index (1). Q

This book presents many double-page reproductions of paintings by Pesek, with additional images from Earth satellites and planetary space missions. The rather brief text provides only a basic introduction for the beginner, but is aided by a number of striking colour diagrams.

Other topics

619 Hodge, P.W.
INTERPLANETARY DUST
New York: Gordon & Breach, 1981, ISBN 0-677-03620-5. 280p; illus, diags. Bibl (3), index (2).

An introductory, though rather technical account, which may interest advanced amateurs. Uses some calculus. Covers the history of the search for interplanetary dust, methods of collection, measurements from Earth satellites, lunar microcraters, deep space measurements, micrometeorites, dynamics, origin of the dust, meteoric dust and the influx of organic molecules, and meteoric dust particles around meteorite craters.

620 Hutchison, Robert
THE SEARCH FOR OUR BEGINNING: AN ENQUIRY BASED ON METEORITE RESEARCH INTO THE ORIGIN OF OUR PLANET AND OF LIFE
London: British Museum (Natural History), 1983, ISBN 0-565-00851-X; and Oxford: Oxford University Press, 1983, ISBN 0-19-858505-5. 164 + 4p; illus (some col.), diags. Glos (2), bibl (1), index (2).

This account of meteoritics is aimed at students, specialists and general readers with a serious interest in the subject. Following a discussion of the debris which arrives on Earth from space, the author turns to meteorites in history and their investigation. The next chapters cover the current state of our geological knowledge of the inner planets, the different varieties of meteorite, and what their study can tell us about the origin of the Solar System in general, and the Earth and life upon it in particular.

*621 LeMaire, T.R.
STONES FROM THE STARS: THE UNRESOLVED MYSTERY OF METEORITES
Englewood Cliffs, NJ: Prentice-Hall, 1980, ISBN 0-13-846881-8. 185 + 4p; illus. Bibl, index.

622 Meadows, Jack
SPACE GARBAGE: COMETS, METEORS AND OTHER SOLAR SYSTEM DEBRIS
London: George Philip, 1985, ISBN 0-540-01087-1. 160p; illus, diags. Glos (2), index (2).

A general account of the material left over from the formation of the Solar System, describing what it is, relationships between the different types, and what its study can tell us about the origin of the Solar System and of life on Earth.

623 Moss, A.A.
METEORITES: A CONCISE ACCOUNT
London: Trustees of the British Museum (Natural History), 1975, ISBN 0-565-05660-3 pbk. 26p; illus, diags. Bibl (1).

This compact work covers the origin of meteorites, their orbits, falls, names, geographical distribution, sizes, craters, morphology, classification and mineral composition, accessory minerals (trace minerals not covered by the classification system), British and Irish meteorites, ages of meteorites, and tektites (pieces of glass of uncertain, but probably extraterrestrial origin). Not listed by Seal (24).

624 Stoneley, Jack, (A.T. Lawton, scientific ed.)
CAULDRON OF HELL: TUNGUSKA
New York: Simon & Schuster, 1977, ISBN 0-671-22943-5. 216 + 8p; illus. Bibl (2), index (2).

A serious, though sensational and sometimes novelistic account of the mysterious, violent impact which occurred at Tunguska in Siberia in 1908. Describes studies of the event and considers possible explanations, including mini-black holes and extraterrestrial spacecraft, as well as more conventional possibilities such as meteorite impact.

The Sun

*625 Brewer, Bryan
ECLIPSE
Ashford, WA: Earth View, Inc., 1978, ISBN 0-932898-26-2 pbk. 102p; illus, diags.

An easy to read, well illustrated account of solar eclipses. Covers past and future eclipses, and provides hints on observing. (From *S&T* May 1979, p 479).

626 Bray, R.J., R.E. Loughhead and C.J. Durant
THE SOLAR GRANULATION (2nd ed.)
Cambridge: Cambridge University Press, 1984, ISBN 0-521-24714-4. 256p; illus, diags. Bibl (3), index (3).

A comprehensive account of observational and theoretical knowledge of the solar granulation (the upper surfaces of bubbles of boiling gas which rise to the solar photosphere), intended for specialists and advanced amateurs.

627 Eddy, John A., (Rein Ise, ed.)
A NEW SUN: THE SOLAR RESULTS FROM SKYLAB
Washington, DC: NASA, 1979, NASA SP-402. 198p; illus & diags (mostly col.). Index (2). Q

A popular account, embellished with many excellent colour images. Provides background information on solar physics, the history of the study of the Sun from space, and the solar telescopes used on Skylab. The results are presented in pictorial form, with descriptive captions.

628 FIRE OF LIFE: THE SMITHSONIAN BOOK OF THE SUN
New York: Norton (Smithsonian Institution), 1981, ISBN 0-89599-

006-7. 263p; illus & diags (mostly col.). Index (2).

This profusely illustrated book is not only concerned with the Sun as an object of interest to astronomers, it is a celebration of our star in its many roles: as an object of worship, as the source of life on Earth, as an inspiration for artists, and as an energy source for mankind.

629 Frazier, Kendrick
OUR TURBULENT SUN
Englewood Cliffs, NJ: Prentice-Hall, 1982, ISBN 0-13-644500-4; 0-13-644492-X pbk. 198 + 24p; illus, diags. Bibl (2), index (2).

A descriptive account for the general reader of our knowledge of the Sun: its origin, evolution, physical characteristics, and its influence on the Earth. Includes some coverage of solar energy.

630 Giovanelli, Ronald
SECRETS OF THE SUN
Cambridge: Cambridge University Press, 1984, ISBN 0-521-25521-X. 116 + 16p; illus (some col.), diags.

This brief, highly illustrated, non-technical account is concerned mainly with the Sun as a star: its movement, features and composition. Includes a chapter on sunspots and the Sun's influence on the weather on Earth.

631 Gribbin, John
THE STRANGEST STAR: A SCIENTIFIC ACCOUNT OF THE LIFE AND DEATH OF THE SUN (UK)/THE DEATH OF THE SUN (US)
London: Athlone Press, 1980, ISBN 0-485-11207-8; Glasgow: Fontana, 1980, ISBN 0-00-635526-9 pbk; New York: Delacorte, 1980, ISBN 0-440-01924-9; New York: Dell (Delta), 1981, ISBN 0-440-51854-7 pbk. 184p; diags. Bibl (2).

This book discusses the importance of the Sun to mankind and to the planet Earth, describes the evolution of a star using the Sun as an example, and considers studies of the Sun which detail its changing patterns of behaviour.

632 Mitton, Simon
DAYTIME STAR: THE STORY OF OUR SUN
London: Faber, 1981, ISBN 0-571-11659-0; New York: Scribner's, 1981, ISBN 0-684-16840-5; 1983, ISBN 0-684-17829-X pbk. 191p; illus, diags. Bibl (1), index (2).

A general, non-technical account of the Sun, including an outline of the historical development of our knowledge, the basic physics of the Sun and its life cycle, modern solar research including the results from Skylab, and solar energy and global energy policy.

633 Nicolson, Iain
THE SUN
London: Mitchell Beazley, 1982, ISBN 0-85533-311-1; New York: Rand McNally, 1982, ISBN 0-528-81540-7. 96p; illus (mostly col.), diags. Glos (1), bibl (1), index (2). Q

One of a series of volumes intended for both amateur and professional astronomers to provide up to date assessments, this book is too advanced for beginners, but does include many good illustrations, with a large colour section in the centre. Covers a wide range of topics in a number of concise chapters, including structure, history of observations, radiation, sunspots, solar flares, the corona, the solar wind, interactions with the Earth, solar energy, oscillations, and the Sun's life cycle. Includes a list of eclipses 1981–2000. (*See also* 600).

634 Noyes, Robert W.
THE SUN, OUR STAR
Cambridge, MA and London: Harvard University Press, 1982, ISBN 0-674-85435-7. 263p; illus, diags. Index (1).

A non-technical introduction to solar studies for the general reader or student. Based on the latest research, particularly that carried out from Skylab, the book covers the Sun's surface, sunspots, the chromosphere, eclipses and the corona, solar flares, the solar wind, the Sun and the Earth's climate, solar energy, and the life history of the Sun. A guide to further reading would have been a useful inclusion. This book is a successor to *Our Sun* (1959) by Donald H. Menzel.

*635 Stafford, Edward
SUN, EARTH AND MAN: THE NEED TO KNOW — THE QUEST FOR KNOWLEDGE OF SUN-EARTH RELATIONSHIPS
Washington, DC: NASA, 1982, NASA EP-172 pbk. 76p; illus.

This colourful booklet describes our quest for information about the relationship between the Earth and the Sun, and the contributions spacecraft are making to that quest. (From *Mercury* May/June 1983, p 91).

636 Washburn, Mark
IN THE LIGHT OF THE SUN: FROM SUNSPOTS TO SOLAR ENERGY
New York and London: Harcourt Brace Jovanovich, 1981, ISBN

0-15-186737-2. 207 + 8p; illus (some col.), diags. Glos (1), bibl (1), index (2).

A popular account of the current state of our knowledge of the Sun. The first half of the book deals with physical aspects, beginning with an account of the birth of the Sun, describing how our knowledge developed. The second half deals with the Sun's effect on the Earth: how it influences life and the climate on our planet. There is also extensive coverage of solar energy, with a chapter on the political factors involved.

637 Zirker, Jack B.
TOTAL ECLIPSES OF THE SUN
New York: Van Nostrand, 1984, ISBN 0-442-29452-7. 210p; illus, diags. Bibl (x), glos (2), index (3).

A book about the science of eclipses, intended for amateur astronomers with a passionate interest in eclipses. It describes the work carried out by professionals over the last decade to show the contribution made by eclipse studies to other astronomical and geophysical studies.

The Inner Planets

638 Adams, Peter
MOON, MARS AND METEORITES
London: HMSO, 1977, ISBN 0-11-880672-6 pbk. 36p; illus & diags (mostly col.). Bibl (1).

A highly illustrated, concise guide for beginners of all ages. Most of this booklet, which concentrates on geological aspects, is devoted to the Moon.

*639 Burgess, Eric
VENUS: AN ERRANT TWIN
New York: Columbia University Press, 1985. 176p; illus, diags.

A survey of our knowledge of Venus, based on observation from Earth and by spacecraft. Describes all the exploratory missions to the planet, from the early American Mariner flights to recent Soviet missions. He considers the 'engineering' of a more hospitable environment on Venus, as well as the possibility of a Venus-like 'runaway greenhouse effect' occurring on Earth. (Publisher's information).

640 Cross, Charles A. and Patrick Moore
THE ATLAS OF MERCURY
London: Mitchell Beazley, 1977, ISBN 0-85533-115-1; New York:

Crown, 1977, ISBN 0-517-53123-2. 48p; illus, diags. Glos (1), bibl (1). Q

A comprehensive survey of Mercury for the amateur astronomer or general reader, produced as a result of data returned by Mariner 10. Includes background information on Mercury in history, early maps, the work of Antoniadi, the astronomy of Mercury, a description of Mariner 10 and its mission, and the mapping of the planet's surface. The charts of Mercury, by Cross, are divided into quadrants and are accompanied by photographic views. There is also some discussion of surface conditions, the atmosphere and the interior.

641 Davies, Merton E., Stephen E. Dwornik, Donald E. Gault and Robert G. Strom
ATLAS OF MERCURY
Washington, DC: NASA, 1978, NASA SP-423. 128p; illus, diags. Bibl (1), index (1). Q

An atlas compiled from images returned by Mariner 10, but covering only the side of the planet which was lighted at the time of the flybys, 1974–75. The shaded relief maps are complemented by many of the photographic images. The introductory chapters provide background information on Mercury, the Mariner 10 spacecraft and its mission, topographic features and surface mapping.

642 Dunne, James A. and Eric Burgess
THE VOYAGE OF MARINER 10: MISSION TO VENUS AND MERCURY
Washington, DC: NASA, 1978, NASA SP-424. 224p; illus (some col.), diags. Bibl (2), index (1). Q

The story of the mission from planning stages to successful completion. Mariner 10 was the first US spacecraft to take photographs of Venus, the first to use the gravity and motion of one planet to alter the path of a spacecraft and send it to another, and the first to explore Mercury. The second half of the book contains four appendices, including 70 pages of images returned by Mariner 10, with explanations and finder maps, and information on the processing of the data returned.

643 Fimmel, Richard O., Lawrence Colin and Eric Burgess
PIONEER VENUS
Washington, DC: NASA, 1983, NASA SP-461 pbk. 253p; illus (mostly col.)., diags. Bibl (3), index (2). Q

The story of the Pioneer Venus programme, covering its history from the planning stages to the successful launch of the orbiter and multiprobe in 1978. Looks at the experiments carried out and the people involved. Includes considerable technical detail, which may prove daunting to the general reader, but features many good colour illustrations.

644 Guest, John, (with Paul Butterworth, John Murray and William O'Donnell)
PLANETARY GEOLOGY
Newton Abbot: David & Charles, 1979, ISBN 0-7135-7739-6. 208p; illus. Index (1).

An introductory work for the amateur astronomer and the general reader. The book is devoted largely to the Moon, Mercury and Mars, but has some coverage of Venus and the satellites of Jupiter. It is arranged in a series of short sections, with the text on the left-hand page and the illustrations on the right.

645 Hunt, Garry E. and Patrick Moore
THE PLANET VENUS
London: Faber, 1982, ISBN 0-571-09050-8. 207p; illus, diags. Bibl (3), indexes (2).

This completely new book, which follows Moore's earlier work of the same title (1956; revised 1963), is a survey of what is now known about Venus, as well as an account of the development of our knowledge of the planet. A scholarly work, with extensive references, but one which should be accessible to the seriously interested amateur or lay reader.

646 Kopal, Zdeněk
THE REALM OF THE TERRESTRIAL PLANETS
London: The Institute of Physics, 1979, ISBN 0-85498-034-2; New York: Halsted Press, 1979, ISBN 0-470-26688-0. 223 + 14p; illus (some col.). Index (2).

This book aims to provide 'an outline of the present extent of our acquaintance with our nearest celestial neighbours'. One brief chapter covers Mercury and distant Pluto; the rest of the book deals with exploration by spacecraft, the Moon, Mars, the asteroids, Venus and Earth.

647 Murray, Bruce, Michael C. Malin and Ronald Greeley
EARTHLIKE PLANETS: SURFACES OF MERCURY, VENUS, EARTH, MOON, MARS.
San Francisco: Freeman, 1981, ISBN 0-7167-1148-6; 0-7167-1149-4 pbk. 387 + 8p; illus (some col.), diags. Bibl (x), index (3).

A textbook intended for 'college undergraduates and . . . other serious non-specialists', which brings 'the subject of the Earthlike planets into a hard intellectual framework, including past parallels and future possibilities'. A non-mathematical treatment which can be used to supplement courses in physical geology, geomorphology, planetary astronomy and planetary science, and vulcanology. The authors' approach is both synthetic and comparative: the Moon, Mercury and Mars have individual chapters, but various features and processes are considered from the point of view of all the bodies in question; for example, chapter three, "Modification from without", covers surface cycles, impacts from space, the effects of gravity and wind. This treatment reflects the changing viewpoint which has developed from the relatively recent ability to investigate the planets at close hand by means of spacecraft.

A detailed account based on data returned by spacecraft is Michael H. Carr's *The geology of the terrestrial planets* (Washington, DC: NASA, 1984, NASA SP-469, 317p).

648 Murray, Bruce and Eric Burgess
FLIGHT TO MERCURY
New York: Columbia University Press, 1977, ISBN 0-231-03996-4. 162p; illus, diags. Index (2). Q

The story of the Mariner 10 mission to Mercury, presented as a chronology of events from the first hints in 1961 to the end of the mission in March 1975. An introductory chapter provides background information on the knowledge of Mercury gained from ground-based observation. Includes many images of the planet's surface.

649 Rükl, Antonín, (Daniela Coxon, trans.)
MOON, MARS AND VENUS
London: Hamlyn, 1976, ISBN 0-600-36219-1. 255p; col. illus, diags. Indexes (3).

Basically a pocket atlas which aims 'to present to the reading public such detail of phenomena pertaining to the Moon, Venus and Mars as can be seen from the Earth'. The introductory chapters, giving background information, are followed by 86 sectional maps of the Moon, each facing a page containing a description of the features shown. This is followed by a map of Mars in six sections. Each of the sets of maps has a separate index. Not listed by Seal (24).

Earth

Atmosphere

650 Akasofu, S.-I.
AURORA BOREALIS: THE AMAZING NORTHERN LIGHTS
Anchorage, AK: Alaska Geographic Society, 1979, ISBN 0-88240-124-6 pbk. 96p; illus (some col.).

A popular introduction by 'one of the world's leading authorities on auroras'. Covers legends of the Northern Lights, the aurora and polar explorers, the aurora and northern settlers, scientific investigation, and how the aurora occurs. Outlines the reasons for auroral study, 'with particular emphasis on its great relevance to our technological future'.

651 Eather, Robert H.
MAJESTIC LIGHTS: THE AURORA IN SCIENCE, HISTORY, AND THE ARTS
Washington, DC: American Geophysical Union, 1980, ISBN 0-87590-215-4. 323p; illus (mostly col.), diags. Bibl (3), indexes (2). Q

A well researched scholarly work intended for all levels of readership. Illustrated with many fine photographs and art reproductions. Includes a chapter describing the author's personal interest in the subject, and an appendix listing the various names for the aurora.

652 Greenler, Robert
RAINBOWS, HALOS, AND GLORIES
Cambridge: Cambridge University Press, 1980, ISBN 0-521-23605-3. 195 + 32p; illus (mostly col.), diags. Bibl (2), index (2).

An account of optical effects for the reader interested in understanding them. Beginning with rainbows, the author moves on to ice-crystal refraction effects, sunlight scattering and diffraction effects, ending with atmospheric refraction (including mirages and the green flash). Some of the excellent colour photographs are used to set puzzles for the reader as part of the author's strategy of encouraging the reader to think about what he or she is seeing. (*See also* 117n).

653 Meinel, Aden and Marjorie
SUNSETS, TWILIGHTS, AND EVENING SKIES
Cambridge: Cambridge University Press, 1983, ISBN 0-521-25220-3.

163 + 24p; illus (mostly col.), diags. Bibl (1), indexes (2).

An account of twilight effects in the atmosphere for the general reader. Coverage includes phenomena such as those associated with sunsets, volcanic eruptions and city lights. Also deals with the green flash, noctilucent clouds, zodiacal light and the gegenschein, twilight on the planets, and effects related to other celestial objects seen from Earth.

654 Young, Louise B.
EARTH'S AURA: A LAYMAN'S GUIDE TO THE ATMOSPHERE
New York: Knopf, 1977, ISBN 0-394-40227-8; New York: Avon, 1979, ISBN 0-380-44230-2 pbk; London: Allen Lane, 1980, ISBN 0-7139-1165-4; Harmondsworth: Penguin (Pelican), 1980, ISBN 0-14-022175-1 pbk. 320p; illus, diags. Bibl (3), index (3).

A survey of our current knowledge of the atmosphere, including the story of mankind's efforts to understand it. Describes and explains simply and clearly phenomena such as the ozone layer, whirlwinds, jet streams and the green flash. Covers also the effects of human interference, the solar wind, cosmic ray bombardment, and the atmospheres of other bodies in the Solar System.

Images from space

655 Bullard, R.K. and R.W. Dixon-Gough
BRITAIN FROM SPACE: AN ATLAS OF LANDSAT IMAGES
London and Philadephia: Taylor & Francis, 1985, ISBN 0-85066-277-X. 128p; illus & diags (mostly col.). Glos (1), index (2). Q

A survey of the whole country presented in 32 images to a uniform scale of 1:500,000, with location maps compiled from satellite images and a brief descriptive text. Also includes eight images at 1:100,000 of special geographic features, and other images illustrating different sensing techniques. The introduction provides information on remote sensing and details of Landsat imaging. Includes a directory of UK organizations associated with remote sensing.

656 Francis, Peter and Pat Jones
IMAGES OF EARTH
London: George Philip, 1984, ISBN 0-540-01083-9; Englewood Cliffs, NJ: Prentice-Hall, 1984; ISBN 0-13-451394-0. 159p; col. illus, diags. Index (2). Q

A collection of astronaut photographs and electronically processed images from Landsat satellites. Includes interpretive text describing the various facets of human influence on the landscape and outlining geographical features.

657 IMAGES OF THE WORLD: AN ATLAS OF SATELLITE IMAGERY AND MAPS
Glasgow: Collins, 1984, ISBN 0-00-360205-2; and Harlow: Longman, 1984, ISBN 0-582-00056-4 (trans. of 1st German ed. 1981). 175p; col. illus, diags. Q

A collection of Landsat images showing selected regions of the world, particularly in Europe, with detailed maps to aid interpretation of what is seen. The images are divided into sections covering different physical features and human activities. An unusual feature for such a book is the modification of the false-colour images to a green base rather than the more common red. The only text in the book is the introductory material, and the interpretive supplement by Roger M. Smith.

658 Sheffield, Charles
EARTH WATCH: A SURVEY OF THE WORLD FROM SPACE
New York: Macmillan, 1981, ISBN 0-02-610090-8; London: Sidgwick & Jackson, 1981, ISBN 0-282-98737-5. 160p; col. illus. Index (1). Q

A collection of Landsat images of the Earth, arranged by geographical feature, e.g. rivers and valleys, cities, deserts, with a final section on 'the signs of man'. Each image is accompanied by a brief descriptive text and an outline locating map. The author emphasizes the value of surveys like this in such fields as agriculture and geology.

659 Sheffield, Charles
MAN ON EARTH: THE MARKS OF MAN: A SURVEY FROM SPACE
New York: Macmillan, 1983, ISBN 0-02-610100-9; London: Sidgwick & Jackson, 1983, ISBN 0-283-98956-4. 160p; col. illus. Index (1). Q

This companion volume to (658) presents a further series of computer-enhanced false-colour images from Landsat spacecraft, most of them previously unpublished. These images show the influence of man on the Earth, covering the ancient world, large engineering projects such as dams, the search for energy and minerals, agriculture, the growth of cities, the changing wilderness. Includes location maps and interpretive text.

660 Short, Nicholas M., Paul D. Lowman, jr., Stanley C. Freden and William A. Finch
MISSION TO EARTH: LANDSAT
Washington, DC: NASA, 1976, NASA SP-360. 459p; illus (mostly col.). Bibl (x), Glos (2), index (2). Q

A huge volume of images of the Earth, with explanatory captions. A separate educator's guide provides a general overview of the mission, describes the operation of the satellite, and explains the techniques of remote sensing and imaging.

Catastrophic changes

661 Asimov, Isaac
A CHOICE OF CATASTROPHES
New York: Simon & Schuster, 1979, ISBN 0-671-22701-7; London: Hutchinson, 1980, ISBN 0-09-141240-4; New York: Fawcett (Columbine), 1981, ISBN 0-449-90048-7 pbk; London: Arrow, 1981, ISBN 0-09-926570-2 pbk. 365p. Index (3).

A survey of the ways in which the world might end. Most of the catastrophes described involve astronomical or geophysical events. They are divided into five classes: the first involves universal changes, such as the reversal of the expansion brought about by the Big Bang; the second involves local cosmic disasters; the third, events rendering the Earth uninhabitable; the fourth, where life on the planet is ended; and the fifth, where human civilization is ended.

662 Clube, Victor and Bill Napier
THE COSMIC SERPENT: A CATASTROPHIST VIEW OF EARTH HISTORY
London: Faber, 1982, ISBN 0-571-11816-X; New York: Universe Books, 1982, ISBN 0-87663-379-3. 299p; illus, diags. Bibl (3), index (2).

The authors bring together 'hitherto unconnected strands in astronomy, biology and geology, and in the early history and mythology of man', to present an unorthodox hypothesis of the importance of external physical influences on Earth history, with emphasis on the importance of comets and their effect on the planet.

663 Gribbin, John and Stephen Plagemann
THE JUPITER EFFECT
London: Macmillan, 1974, ISBN 0-333-17418-6; New York: Random House (Vintage Books), 1976, ISBN 0-394-72221-3 pbk; London:

Fontana, 1977, ISBN 0-00-634419-4 pbk. 136p; diags. Index (1).

A book well known for the author's prediction that the 1982 near alignment of the planets as seen from Earth would lead to increased solar activity which would 'trigger off regions of earthquake activity on Earth . . . the evidence . . . points to 1982 as the year in which the Los Angeles region of the San Andreas fault will be subjected to the most massive earthquake known in the populated regions of the Earth in this century'. (*See also* 664).

664 Gribbin, John and Stephen Plagemann
THE JUPITER EFFECT RECONSIDERED (US)/BEYOND THE JUPITER EFFECT (UK)
New York: Random House (Vintage Books), 1982, ISBN 0-394-70827-X pbk; London and Sydney: Macdonald, 1983, ISBN 0-356-08686-0. 184p. Bibl (x), index (2).

A follow-up to (663) in which the authors admit they were wrong. The Sun reached a peak of activity in 1979 rather than 1982, and did not trigger the predicted earthquakes. This book takes a broad view in attempting to combat the extreme reactions to the failure of the authors' original hypothesis. They discuss the complex forces which act on the Earth from within and without to determine geological and climatic processes.

665 Moore, Patrick
COUNTDOWN! OR HOW NIGH IS THE END?
London: Michael Joseph (Rainbird), 1983, ISBN 0-7181-2291-7. 136 + 24p; illus, diags. Index (2).

An often light-hearted look at ideas and hypotheses about the end of the world, from the pseudo-scientific to serious astronomical speculation.

*666 Morris, Richard
THE END OF THE WORLD
Garden City, NY: Doubleday (Anchor Press), 1980, ISBN 0-385-15523-9. 180p; illus. Bibl, index.

667 Warlow, Peter
THE REVERSING EARTH
London: Dent, 1982, ISBN 0-460-04478-8. 213p; diags. Bibl (3), index (2).

The author considers the possibility of a reversal of the Earth's rotation, and an inversion of the planet, as explanations for catastrophes such as the ice ages, the biblical flood, reversals of the Earth's magnetic field, and sudden changes of climate and culture.

This highly speculative book ends with a new view of the formation of the planets.

The Moon

668 Alter, Dinsmore, (revised by Joseph H. Jackson)
PICTORIAL GUIDE TO THE MOON (3rd ed.)
New York: Crowell, 1979, ISBN 0-690-01824-X pbk (updated reprint of 3rd ed. 1973). 224p; illus, diags. Glos (2), index (2). Q

An introduction for the general reader and amateur astronomer which contains many photographs taken from Earth, as well as images from space. Provides a useful guide for the identification of lunar features.

669 Beckdale, J.
EXPLORING: MAN ON THE MOON
London: HMSO, 1977, ISBN 0-11-290275-8 pbk. 28p; illus (some col.), diags. Bibl (1).

A concise guide, for beginners of all ages, to manned and unmanned exploration of the Moon, produced to mark the opening of the Exploration exhibition at the Science Museum in London, 1977.

670 Cadogan, Peter H.
THE MOON: OUR SISTER PLANET
Cambridge: Cambridge University Press, 1981, ISBN 0-521-23684-3; 0-521-28152-0 pbk. 391p; illus, diags. Glos (2), bibl (2), index (1).

A survey of, and introductory text to lunar science. Non-mathematical, but too detailed for the general reader. The six parts of the book cover 1. the Moon from a distance (its motion, its place in history, spacecraft flybys, lunar geology, craters), 2. Moon landers, 3. Moon rocks and minerals, 4. the regolith (lunar 'soil'), 5. global measurements, and 6. the Moon's history and future.

671 French, Bevan M.
THE MOON BOOK
Harmondsworth: Penguin, 1977, ISBN 0-14-004340-3 pbk. 287 + 16p; illus, diags. Bibl (1), index (3).

A survey of our knowledge of the Moon for the seriously interested lay reader, written by a geologist. Begins with the arrival of the first Moon rocks on Earth, carried by the returning Apollo 11 spacecraft in 1969, then goes on to describe briefly the history of observation of the Moon, and unmanned and manned spacecraft investigation, before considering what we know about the origin of the Moon, its composition, its surface and interior. Ends with a look at the wider implications of the exploration of the Moon and the knowledge gained therefrom.

672 Guest, J. and R. Greeley
GEOLOGY ON THE MOON
London: Wykeham, 1977, ISBN 0-85109-580-1; 0-85109-540-2 pbk; New York: Crane, Russak, 1977, ISBN 0-8448-1170-X. 235p; illus, diags. Bibl (3), indexes (2).

Part of a series of introductory works for prospective undergraduates, this book is concerned with what has been learned over the past fifteen years (up to 1977) about the processes of meteoritic impact and volcanism which have produced the present face of the Moon, particularly from the intensive study of returned samples. Emphasizes, necessarily, the photogeological approach, which so far is the only feasible geological technique for the overall study of the planets beyond Earth.

673 Manson, Lewis A.
THE BIRTH OF THE MOON
Santa Monica, CA: Dennis-Landman, 1978, ISBN 0-930422-12-0. 205p; illus & diags (some col.). Bibl (x), index (1). Q

A personal account of the formation of the Moon, which not only contradicts established theories, but also ignores evidence which would disprove the author's hypothesis.

674 Masursky, Harold, G.W. Colton and Farouk El-baz (ed.)
APOLLO OVER THE MOON: A VIEW FROM ORBIT
Washington, DC: NASA, 1978, NASA SP-362. 255p; illus & diags (some col.). Glos (2), bibl (1). Q

This well produced book contains a selection of lunar views from various Apollo missions, accompanied by substantial descriptive captions. Includes geological maps of particular areas in colour. An introductory chapter describes the selection and use of photographic systems, the formulation of photographic mission plans, analysis of data, and some scientific results.

675 Moore, Patrick
GUIDE TO THE MOON (UK)/THE NEW GUIDE TO THE MOON (US)
Guildford: Lutterworth Press, 1976, ISBN 0-7188-1720-6; New York: Norton, 1977, ISBN 0-393-06414-X. 320 + 16p; illus, diags. Bibl (1),

indexes (3).

A general introduction written from an observer's point of view; the author presents his own opinions on controversial topics. Coverage includes the Moon's origin, its motion, its relationship to Earth, a history of observations, physical features, eclipses, and exploration. The appendices cover observing the Moon, numerical data, eclipses of the Moon 1976–86, and lunar landings.

676 Moore, Patrick
THE MOON
London: Mitchell Beazley, 1981, ISBN 0-85533-310-3; New York: Rand McNally, 1981, ISBN 0-528-81541-5. 96p; illus (mostly col.), diags. Glos (1), bibl (1), index (1). Q

One of a series of volumes all produced in the same format, this survey of lunar science covers a wide range of topics in two-page spreads: the Moon's origin, orbit, eclipses and tides, history of observations, unmanned probes, manned flights, physical features, interior, surface, maria, highlands, craters, unusual features, and cartography. Includes a series of maps produced by Charles A. Cross, and a list of lunar probes 1958–80, as well as notes on observing. (*See also* 600).

677 Whitcomb, John C. and Donald B. de Young
THE MOON: ITS CREATION, FORM AND SIGNIFICANCE
Winona Lake, IN: BMH Books, 1978, ISBN 0-8010-9619-7. 180 + 16p; illus (some col.), diags. Bibl (2), indexes (2).

A creationist view of the Moon, by a professor of theology and a physicist.

Mars

678 Baker, Victor R.
THE CHANNELS OF MARS
Austin, TX: University of Texas Press, 1982, ISBN 0-292-71068-2; Bristol: Adam Hilger, 1982, ISBN 0-85274-467-6. 198p; illus, diags. Bibl (2), index (2). Q

A survey of the geomorphology of Mars, focusing on the channels, which have been proved not to be canals, but still present a puzzle. Probably too specialized for the general reader, but should appeal to someone with a strong interest in Mars or geology. Describes what is known about the landforms and physical processes of Mars, based mainly on the author's own research. Includes chapter summaries.

679 Batson, R.M., P.M. Bridges and J.L. Inge
ATLAS OF MARS
Washington, DC: NASA, 1979, NASA SP-438. 146p; illus, diags. Bibl (2), indexes (3). Q

This book 'comprises small-scale [1:5,000,000] maps and photomosaics covering the entire surface of the planet', with data on the photographs incorporated. Includes a 15-page gazetteer of names used for the features, with a note on the derivation of each. Not for casual reading.

680 Blunck, Jürgen
MARS AND ITS SATELLITES: A DETAILED COMMENTARY ON THE NOMENCLATURE (2nd ed.)
Smithtown, NY: Exposition, 1982, ISBN 0-682-49777-0. 222p; diags. Index (3).

The product of thorough research into the origins of names used for the features of Mars. The names, with notes, are listed in a series of alphabetically arranged sectional gazetteers. Appendices cover older nomenclature, rarely used surface designations, maps of the ancient world, science fiction names and nicknames, named rocks at the Viking 1 and 2 lander sites, IAU committees on Martian nomenclature, and a short list of maps and globes of Mars.

681 Burgess, Eric
TO THE RED PLANET
New York: Columbia University Press, 1978, ISBN 0-231-04392-9. 181p; illus, diags. Index (3). Q

A personal account of the exploration of Mars by a science writer who first saw the planet through his telescope in 1934, and has been writing about it since 1936. He deals with the history of observations, early space missions, the Viking mission and the resulting discoveries, in an account enlivened by countless reports of conversations with the people involved with various aspects of this endeavour.

682 Carr, Michael H. and Nancy Evans
IMAGES OF MARS: THE VIKING EXTENDED MISSION
Washington, DC: NASA, 1980, NASA SP-444 pbk. 32p; illus. Q

Images from the two Viking orbiters, launched in 1976, which continued to operate long after original expectations. Covers a wide range of surface features, and includes a few lander images.

683 Carr, Michael H.
THE SURFACE OF MARS
New Haven, CT and London: Yale University Press, 1981, ISBN 0-300-02750-8; 1984, 0-300-03242-0 pbk. 332p; illus, diags. Bibl (3), index (2). Q

A comprehensive survey of our knowledge of Mars, concentrating on the geology, by the leader of the Viking Orbiter Imaging Team. Requires a knowledge of physics and chemistry, along with the basic vocabulary of geology. Includes an essay, "The search for life on Mars" by Harold P. Klein.

684 Chandler, David L.
LIFE ON MARS
New York: Dutton, 1979, ISBN 0-525-14560-5. 212p; illus, diags. Bibl (1), index (2).

A popular account, which provides background information on life and evolution, and Mars, before dealing with the Viking mission and the search of life. Draws a comparison with the Earth in considering the problems of searching for life on a planet from space, and looks at the possibility of future Martian searches.

685 Cooper, Henry S.F.
THE SEARCH FOR LIFE ON MARS: EVOLUTION OF AN IDEA
New York: Holt, 1980, ISBN 0-03-046166-9; 1981, ISBN 0-03-059818-4 pbk. 254p. Index (2).

The author ties the story of the search for life on Mars carried out with the Viking landers to the personalities involved. The first half of the book is devoted to Carl Sagan, one of the chief proponents of the universality of life, and to the preparation of the Viking mission. The second half deals with the analysis of the results, focusing on the biologists, Harold P. Klein, Norman Horowitz, Gilbert Levin and Vance Oyama. Most of this book is based on material which originally appeared in the *New Yorker* magazine.

686 Firsoff, V.A.
THE NEW FACE OF MARS
Hornchurch: Ian Henry, 1980, ISBN 0-86025-818-1. 164p; illus, diags. Bibl (1), index (1).

A survey of our knowledge of Mars, 'accessible to a reader without specialized knowledge, but it is not elementary and does not skip over difficulties'. Not an easy read for the beginner, the book describes the Viking missions as well as previous observations and space probes. Coverage includes geology, physical features, climate, and the moons.

687 French, Bevan M.
MARS: THE VIKING DISCOVERIES
Washington, DC: NASA, 1977, NASA EP-146 pbk. 36p; illus (some col.). Bibl (1). Q

A brief, illustrated summary of the Viking mission and its results.

688 Lowell, Percival
MARS
Bernardstown, MA: Paul W. Luther (Astronomy Books), 1978, (reprint of 1st ed. 1895). 228p; illus (1 col.), diags. Indexes (3).

A facsimile reprint of Lowell's popular classic in which he presented his ideas about the red planet, notably that it was inhabited by intelligent life, a theory based on his belief in the supposed canals.

689 MARS AS VIEWED BY MARINER 9: A PICTORIAL PRESENTATION BY THE MARINER 9 TELEVISION TEAM AND THE PLANETOLOGY PROGRAM PRINCIPAL INVESTIGATORS
Washington, DC: NASA, 1979, NASA SP-329 (revised reprint of 1st ed. 1976). 225p; illus. Q

A selection of the more than 7,300 images sent back to Earth by Mariner 9 in 1972. Arranged by feature, e.g. giant volcanoes, channels, fractures and faults, craters, wind-shaped features, polar regions, clouds, moons and enigmas.

690 Moore, Patrick
GUIDE TO MARS
Guildford: Lutterworth Press, 1977, ISBN 0-7188-2316-8; New York: Norton, 1977, ISBN 0-393-06432-8. 214 + 8p; illus, diags. Indexes (2).

A survey of our knowledge of Mars, with a strong historical emphasis, covering observation and exploration: the discovery of the various surface and physical features, including the story of the 'canals', and the results of the Viking missions. One of the five appendices provides notes on observing Mars.

*691 Rovin, Jeff
MARS!
Los Angeles: Pinnacle (Corwin Books), 1978, ISBN 0-89474-011-3. 244p; illus. Index.

Starting with fact and early astronomy, the author then moves to Martians in fiction. The second half of the book fully describes and illustrates the Viking landings. (From Blashfield (15), p 102).

692 Spitzer, Cary R. (ed.) and the Viking Orbiter Imaging Team
VIKING ORBITER VIEWS OF MARS
Washington, DC: NASA, 1980, NASA SP-441. 182p; illus (some col.). Glos (2), bibl (1). Q

A selection of the more than 46,000 images sent back by the highly successful Viking orbiter (up to May 1980). Arranged in sections covering different features, e.g. channels, volcanic features, craters, moons, surface processes, polar regions. Each image is accompanied by a short descriptive caption. A collapsible viewer for the small selection of stereoscopic images is included in a pocket inside the back cover.

693 Viking Lander Imaging Team
THE MARTIAN LANDSCAPE
Washington, DC: NASA, 1978, NASA SP-425. 160p; illus (some col.), diags. Q

The introductory chapters, with the overall title of "An anecdotal account", by Thomas A. Mutch, tell the story of the preparation of the mission. This adds an extra dimension to the selection of images from the two Viking landers, which are accompanied by a brief, descriptive text. The final section of images is a set of 'special effects' pictures. Includes a collapsible viewer for the small selection of stereoscopic images in a pocket inside the back cover.

694 Washburn, Mark
MARS AT LAST!
New York: Putnam, 1977, ISBN 0-399-11935-3; London: Sphere (Abacus), 1979, ISBN 0-349-13591-6 pbk. 243 + 8p; col. illus. Bibl (1), index (3).

The story of mankind's involvement with the red planet, from mythology to unmanned landing. Covers the development of our knowledge from the earliest observations, through the story of the 'canals', to the advances made possible by orbiting missions, and finally the Viking landers. The author looks also at the image of Mars in fiction, and discusses unorthodox theories relating to the planet.

The Outer Planets

General

695 Alexander, A.F.O'D.
THE PLANET SATURN: A HISTORY OF OBSERVATION, THEORY AND DISCOVERY
New York: Dover, 1980, ISBN 0-486-23927-6 pbk (corrected reprint of 1st ed. 1962). 474 + 23p; illus, diags. Bibl (x), index (3).

The classic account of the development of our knowledge of Saturn before the space age. A comprehensive treatment of observations from ancient times to the mid-twentieth century; rigorous, but non-mathematical.

For a similar treatment of the seventh planet by the same author, *see The planet Uranus* (London: Faber, 1965).

696 Batson, Raymond A.
VOYAGER 1 AND 2 ATLAS OF SIX SATURNIAN SATELLITES
Washington, DC: NASA, 1984, NASA SP-474. 175p; illus (some col.), diags. Bibl (1). Q

Images of Mimas, Enceladus, Tethys, Dione, Rhea and Iapetus, each in two versions: contrast-enhanced unfiltered and contrast-enhanced high pass filtered, and maps based upon them in different projections for the same area. Appendices provide details of the Voyager mission, the Saturnian system, image processing, mapping and feature nomenclature.

697 Elliot, James and Richard Kerr
RINGS: DISCOVERIES FROM GALILEO TO VOYAGER
Cambridge, MA: MIT Press, 1984, ISBN 0-262-05031-5. 209p; illus, diags. Bibl (2), index (2).

A popular account of the discoveries of planetary rings, by a space scientist with first-hand involvement (Elliot) and a reporter, which includes sufficient technical detail to properly trace the paths to discovery, and convey some of the wonder of planetary rings. The authors' aim is to tell a story that is not found in the textbooks: the accidental and essentially unscientific way in which the rings were found, showing the complex and circuitous nature of scientific discovery.

698 Gehrels, Tom and Mildred Shapley Matthews (ed.)
SATURN
Tucson, AZ: University of Arizona Press, 1984. ISBN 0-8165-0829-1.

968p; illus (some col.), diags. Bibl (3), glos (3), index (3).

This volume in the publisher's 'Space Science Series' of comprehensive overviews for the specialist is less technical than most of the others and may appeal to the seriously interested amateur. The articles are divided into seven parts: 1. introduction (overview and brief historical survey), 2. Saturn (structure and atmosphere), 3. magnetosphere, 4. rings, 5. satellites, 6. Titan, 7. origin.

Other volumes in this series cover asteroids, comets (*see* 724), Jupiter, planetary rings, planetary satellites, planets stars and nebulae, satellites of Jupiter, and Venus.

699 Hunt, Garry and Patrick Moore
JUPITER
London: Mitchell Beazley, 1981, ISBN 0-85533-309-X; New York: Rand McNally, 1981, ISBN 0-528-81542-3. 96p; illus (mostly col.), diags. Glos (1), bibl (1), index (1). Q

One of a series of volumes all produced in the same format, this one incorporates data returned by the Pioneer and Voyager spacecraft to present an up-to-date survey of our knowledge of Jupiter and its satellites. Coverage includes nomenclature, history of observations, the Voyager spacecraft, the structure of the recently discovered ring, maps of the four Galilean satellites. Includes position tables and notes on observing Jupiter. (*See also* 600).

700 Hunt, Garry and Patrick Moore
SATURN
London: Mitchell Beazley, 1982, ISBN 0-85533-354-5; New York: Rand McNally, 1982, ISBN 0-528-81545-8. 96p; illus (mostly col.), diags. Glos (1), bibl (1), index (1). Q

One of a series of volumes all produced in the same format, this one incorporates data returned by the Pioneer and Voyager spacecraft to present an up-to-date survey of our knowledge of Saturn. Coverage includes motion, history of observations, the Voyager spacecraft, structure, the rings, and satellites. Includes position tables 1982–92 and notes for the observer. (*See also* 600).

701 Hunt, Garry
URANUS AND THE OUTER PLANETS
Cambridge: Cambridge University Press, 1982, ISBN 0-521-24573-7. 307p; illus, diags. Index (1).

The proceedings of the IAU/RAS Colloquium 60, a special meeting held to celebrate the 200th anniversary of the discovery of Uranus in 1781. Not a book for the lay reader, but may interest the keen amateur. Listed here chiefly for its section on the history of the discovery of the planet, with a facsimile of Herschel's original article on the subject. The remainder of the papers deal with the present state of knowledge, and are mostly about Uranus.

702 Peek, Bertrand M., (revised by Patrick Moore)
THE PLANET JUPITER (2nd ed.)
London: Faber, 1981, ISBN 0-571-18026-4. 240 + 14p; illus, diags. Index (1).

A book which provides background information for the serious amateur astronomer. The outdated theoretical section from the first edition has been omitted, but the observational chapters have been left unchanged from their form of 1958, because of the prohibitive increase in cost that would have been incurred in updating, which, in any case, 'would not have added so much to [the book's] intrinsic value as might be thought'. The information on the satellites has been updated. The bulk of the book is devoted to an account of the observable features of Jupiter's surface.

703 Whyte, A.J.
THE PLANET PLUTO
Toronto: Pergamon, 1980, ISBN 0-08-024648-6. 147p; illus, diags. Bibl (3), indexes (2).

A scholarly account of the search for and discovery of Pluto, and subsequent developments in our understanding of the planet. Not for casual reading. Ends by speculating on a possible future orbital mission to Pluto. (*See also* 410 and 414).

Exploration

704 Burgess, Eric
BY JUPITER: ODYSSEYS TO A GIANT
New York: Columbia University Press, 1982, ISBN 0-231-05176-X. 155p; illus, diags. Index (2). Q

The story of the exploration of Jupiter, with emphasis on the Pioneer and Voyager missions. Describes the planning stages, the experiments carried out, and discusses the scientific results. Like the author's book about the exploration of Mars (681), this one includes comments from people such as project scientists and managers to emphasize the human aspects of the venture.

705 Cooper, Henry S.F.
IMAGING SATURN
New York: Holt, 1983, ISBN 0-03-061688-3. 210 + 8p; illus (some

col.), diags.

A journalistic account of the Voyager missions to Saturn, told largely through interviews with the Voyager Imaging Team.

706 Fimmel, Richard O., James van Allen and Eric Burgess
PIONEER: FIRST TO JUPITER, SATURN, AND BEYOND
Washington, DC: NASA, 1980, NASA SP-446 pbk. 285p; illus (some col.), diags. Index (2). Q

A mostly non-technical account of the Pioneer missions, from the planning stages to the discoveries made, which also provides background information on Jupiter and Saturn. Includes excellent illustrations of all stages of the Pioneer 10 and 11 missions, which may be enjoyed by anyone, but the text is not for beginners. A separate bibliography was published as NASA TM-81233. (This is an updated version of NASA SP-349 and SP-369).

707 Morrison, David and Jane Samz
VOYAGE TO JUPITER
Washington, DC: NASA, 1980, NASA SP-439 pbk. 199p; illus & diags (mostly col.). Bibl (1).

A popular account of the Voyager missions to Jupiter, with background details and a discussion of the findings. Profusely illustrated and told in a readable style.

708 Morrison, David
VOYAGES TO SATURN
Washington, DC: NASA, 1982, NASA SP-451 pbk. 227p; illus (some col.). Bibl (1), glos (1), index (1).

A readable, profusely illustrated account of the Voyager 1 and 2 encounters with Saturn, in 1980 and 1981 respectively. Describes the mission and summarizes the findings.

709 NASA Ames Research Center
PIONEER SATURN ENCOUNTER
Moffett Field, CA: NASA Ames Research Center, 1979, pbk. 28p; illus & diags (mostly col.). Q

A selection of images returned by Pioneer 11, with a brief text describing them, and noting 'science highlights'.

710 NASA
VOYAGER ENCOUNTERS JUPITER
Washington, DC: NASA, 1979, pbk. 40p; illus & diags (mostly col.). Q

Published to provide an early look at the Jovian planetary system as photographed by Voyagers 1 and 2, which were launched in the summer of 1977, and passed Jupiter in 1979. An excellent pictorial presentation, with a brief descriptive text.

711 NASA/Jet Propulsion Laboratory
VOYAGER FLIGHTS TO JUPITER AND SATURN
Washington, DC: NASA, 1982, NASA EP-191 pbk. 61p; illus (mostly col.). Q

A highly illustrated popular account of the Voyager missions, with a brief summary of the findings.

712 NASA
VOYAGER 1 ENCOUNTERS SATURN
Washington, DC: NASA, 1980, pbk. 40p; illus & diags (mostly col.). Q

Published to provide an early look at the images returned by Voyager 1, this is an excellent pictorial presentation, with short descriptive captions, and notes on 'scientific highlights' and 'scientific investigations'.

713 NASA/Jet Propulsion Laboratory
VOYAGER AT SATURN: 1981
Pasadena, CA: Jet Propulsion Laboratory, 1981, pbk. 16p; col. illus & diags. Q

Produced just before the Voyager 2 encounter with Saturn, this booklet presents colour images from Voyager 1, and provides background information on the Voyager missions and the Saturnian system.

714 Washburn, Mark
DISTANT ENCOUNTERS: THE EXPLORATION OF JUPITER AND SATURN
San Diego, CA: Harcourt Brace Jovanovich, 1983, ISBN 0-15-125744-2; 0-15-626108-1 pbk. 272 + 8p; illus (some col.), diags. Index (2).

A readable popular account of the Voyager missions, written by an outside observer, but based on three years of research, during which the author interviewed many of the people involved. He hopes to convey the idea that through distant radio links we can all participate in such ventures. Describes the state of our knowledge of Jupiter and Saturn before the missions, the build up to the launch, and the findings.

Comets

General

715 Brandt, John C.
COMETS: READINGS FROM *SCIENTIFIC AMERICAN*
San Francisco: Freeman, 1981, ISBN 0-7167-1319-5; 0-7167-1320-9 pbk. 92p; illus & diags (some col.). Bibl (1), index (1). Q

A collection of ten varied items, prefaced by "The astronomy of comets" by Brandt. Reflecting the way comets have captured the non-scientist's imagination, the first article is "Giotto's portrait of Halley's comet". This is followed by facsimiles of *Scientific American* reports of the 1910 appearance of Halley's comet. The other articles, published from 1951 onward, show the development of our knowledge of comets, and the wider implications of their study, for example in understanding the solar wind, and the origin of the Solar System.

*716 Brandt, John C.
COMETS AND COMET HALLEY
San Francisco: Astronomical Society of the Pacific, 1984, boxed. 16p; + 31 slides (17 col.).

Compiled by a space scientist, these highly informative illustrations cover the physical nature of comets, how they have been regarded throughout history, and the observing prospects for Halley's Comet. The slides and booklet combine to deliver a remarkably complete introduction to the subject. (From *S&T* June 1985, p 519). (Address: *see* 139).

717 Brandt, John C. and Robert D. Chapman
INTRODUCTION TO COMETS
Cambridge: Cambridge University Press, 1981, ISBN 0-521-23906-0; 1982, ISBN 0-521-27218-1 pbk. 246p; illus, diags. Bibl (3), indexes (2).

A thorough introductory work, which requires some background knowledge on the part of the reader. It consolidates 'in one text information from the fields of history, physics, astronomy and space science'. Of the book's four sections, the first is historical, the second deals with current knowledge of structure, dynamics and origins, the third looks at recent developments and possible future space missions, while the fourth examines the public fascination with comets.

718 Chapman, Robert D. and John C. Brandt
THE COMET BOOK: A GUIDE FOR THE RETURN OF HALLEY'S COMET
Boston: Jones & Bartlett, 1984, ISBN 0-86720-029-4 pbk. 168 + 4p; illus (some col.). diags. Bibl (1), glos (2), index (2).

A popular general account of comets, largely from a historical point of view, with a special section on the 1985–86 return of Halley's Comet, including charts and notes to aid in locating it. The book begins with a description of the arrival of a comet, its appearance and physical properties.

*719 James, Peter
COMETS AND CATASTROPHES
London: Severn House, 1985, ISBN 0-7278-2070-2; 0-7278-2069-9 pbk. 160p; illus.

A popular review of comet science and superstition from prehistoric times to the present. (Publisher's information).

720 Moore, Patrick
GUIDE TO COMETS (2nd ed.)
Guildford: Lutterworth Press, 1977, ISBN 0-7188-2315-X; 111 + 16p; illus, diags. Glos (1), index (2).

A general introduction, which combines historical matter with a survey of current knowledge in dealing with the paths of comets, comet hunters, Halley's Comet, lost comets, and where comets come from. (Originally published as *The comets: visitors from space*).

*721 Ritchie, David
COMETS: THE SWORDS OF HEAVEN
New American Library (Plume), 1985, ISBN 0-452-25619-4 pbk. 256p.

A comprehensive, popular introduction to comets in general, with particular attention paid to Halley's Comet. Following discussions of comet lore and the development of cometary science, the author embarks the reader on a fanciful ride on a comet's nucleus, inward through the asteroid belt, to a rendezvous with the Sun. Also considers effects related to cometary impacts upon Earth, and comets in mythology and the Bible. Includes a guide for observing Comet Halley and describes orbital missions to the comet. (From *S&T* Aug. 1985. p 132).

*722 Sagan, Carl and Ann Druyan
COMET
New York: Random House, 1985, ISBN 0-394-54908-2; London:

Michael Joseph, 1985, ISBN 0-7181-2631-9. 240p; illus. Index.

An introduction to comets in general and Halley's Comet in particular. (Publisher's information).

723 Seargeant, David A.
COMETS: VAGABONDS OF SPACE
New York: Doubleday, 1982, ISBN 0-385-17869-7. 234p; illus, diags. Bibl (x), glos (1), index (3).

This book for the serious amateur bridges the gap between popular introductory works and specialist texts. The four sections cover the physical make-up of comets, their discovery and orbits, famous comets, and the most famous of all, Halley's. Appendices deal with observational techniques and basic data reduction, and historical comets bright enough to have become visible in broad daylight.

724 Wilkening, Laurel L. (ed.), (assisted by M.S. Matthews)
COMETS
Tucson, AZ: University of Arizona Press, 1982, ISBN 0-8165-0769-4. 766p; illus, diags. Bibl (x), glos (3), index (2).

This volume in the publisher's 'Space Science Series' (*see* 698) of comprehensive overviews for the specialist is also intended for the serious amateur. The review papers by 48 specialists are divided into seven sections: 1. overview, 2. nucleus, 3. dust, 4. coma, 5. ion tails and solar wind interactions, 6. origin, evolution and interrelations, 7. appendices.

Halley's Comet

*725 Asimov, Isaac
ASIMOV'S GUIDE TO HALLEY'S COMET: THE AWESOME STORY OF COMETS
New York: Walker, 1985, ISBN 0-317-13623-2; London: Granada, 1985, ISBN 0-586-06685-3 pbk. 118p; diags. Index.

A simplified introduction to comets — only about 20 pages deal directly with Halley's Comet — which balances astronomy, history and celestial mechanics with an interesting blend of superstition and comet lore. (From *Astronomy* July 1985, p 56).

726 Calder, Nigel
THE COMET IS COMING: THE FEVERISH LEGACY OF MR. HALLEY
London: BBC, 1980, ISBN 0-563-17859-0; New York: Viking, 1981, ISBN 0-670-23216-5; Harmondsworth: Penguin, 1982, ISBN 0-14-

006069-3 pbk. 160 + 8p; illus (some col.), diags. Bibl (2), index (1).

A well illustrated account of historical sightings of Halley's Comet, with details of associated prophecies of doom, and an account of phenomena supposedly linked with comets, including flu epidemics and the death of the dinosaurs.

*727 Freitag, Ruth S.
HALLEY'S COMET: A BIBLIOGRAPHY
Washington, DC: Library of Congress, 1984, ISBN 0-8444-0459-4. 25 + 555p; illus. Indexes.

An extensive, annotated bibliography listing 3,235 items from broadsides to books, including fact and fiction, arranged alphabetically. (From *S&T* June 1985, p 519).

728 Harpur, Brian
THE OFFICIAL HALLEY'S COMET BOOK
London: Hodder & Stoughton, 1985, ISBN 0-340-36511-0. 184 + 8p; illus (some col.), diags. Glos (2), index (2).

This book, intended as the official handbook of the Halley's Comet Society, is 'designed to be a complete guide to all the facts, myths and legend which surround the visitations of the comet' since its first recorded appearance in 240 BC to date. Coverage includes the 1910 appearance, its commercial exploitation, and verse inspired by it; astronomy up to Halley; Halley's life; comets in literature and history; and advice on where to look. Includes a biographical A-Z of selected astronomers.

729 Lancaster-Brown, Peter
HALLEY AND HIS COMET
Poole: Blandford, 1985, ISBN 0-7137-1447-6. 186 + 8p; illus, diags. Bibl (2), index (2).

A largely historical account of the famous comet, and the role of the man after whom it was named. Details Halley's association with Isaac Newton, who solved the problem of cometary motion, leading to Halley's prediction that a comet he had seen in 1682 would return in 1758. Also describes more recent studies of comets, comets and life (the ideas of Hoyle and Wickramasinghe, *see* 499–506), and missions to rendezvous with Halley's Comet in 1985–86. Appendices cover the planned space missions, comet nomenclature, cometary orbits, magnitude or brightness.

Two biographies of Halley are: Angus Armitage, *Edmund Halley* (London: Thomas Nelson, 1966, 220p), and Colin A. Ronan, *Edmund Halley: genius in eclipse* (London: Macdonald, 1969, ISBN 0-356-02942-5; Garden City, NY: Doubleday, 1969, 251p).

*730 Metz, Jerred
HALLEY'S COMET 1910: FIRE IN THE SKY
St. Louis, MO: Singing Bone Press, 1985, ISBN 0-933439-00-8; 0-933439-01-6 pbk. 124p; illus.

Not a book about Halley's Comet itself, nor even an astronomy book, but one in which the public interest shown in the 1910 return is recreated in the form of cartoons, advertisements, postcards and the news and views of the time — all written and presented in light-hearted fashion. (From *Spaceflight* Sept./Oct. 1985, p 359).

731 Moore, Patrick and John Mason
THE RETURN OF HALLEY'S COMET
Cambridge: Patrick Stephens, 1984. ISBN 0-85059-667-X; New York: Norton, 1984, ISBN 0-393-01827-5; Wellingborough: Patrick Stephens, 1985, ISBN 0-85059-790-0 pbk; New York: Warner, 1985, ISBN 0-446-38303-1 pbk. 121p; illus, diags. Index (2).

This book provides a sound general introduction to comets in general, from a largely historical point of view, in addition to its coverage of Comet Halley. Includes charts and details of the 1985–86 visit. (The paperback edition has been updated and corrected).

732 Morton, Bruce
HALLEY'S COMET, 1755–1984: A BIBLIOGRAPHY
Westport, CT: Greenwood Press, 1985, ISBN 0-313-24022-1. 280p; illus. Glos (1), indexes (3).

Intended to explore the social and cultural aspects of Halley's Comet, and not aimed primarily at scientists, this selective, annotated bibliography contains 1,301 entries, mostly covering periodical articles in the English language. Includes comprehensive coverage of the *New York Times* and the London *Times*. Chronologically arranged, the book lists material published from 1755 to April 1984.

733 Reddy, Francis
HALLEY'S COMET!
Milwaukee, WI: AstroMedia, 1985, ISBN 0-913135-02-X; London: Pan, 1985, ISBN 0-330-29164-5 spiral-bound. 59p; illus & diags (mostly col.). Bibl (1). Q

This colourful guide includes background information on comets in general, and on previous visits of Comet Halley, in its brief text. Includes 20 finder charts giving very clear guidance for the novice, and a cut-out-and-assemble model of the comet's orbit. The UK edition is accompanied by a poster showing the path of the comet and selected scenes from history.

*734 Stephenson, F.R. and C.B.F. Walker (ed.)
HALLEY'S COMET IN HISTORY
London: British Museum, 1985, ISBN 0-7141-1118-X pbk. 64p; illus, diags.

Reveals the first known record of the sighting of the comet at its return of 164 BC on a Babylonian tablet, and discusses Comet Halley's importance in the history of astronomy. (Publisher's information).

735 Tattersfield, D.
HALLEY'S COMET
Oxford: Blackwell, 1984, ISBN 0-631-13558-8. 164p; illus, diags. Bibl (2), index (2).

A general introduction covering most aspects of Halley's Comet: history, composition, motion, observing and photographing it. Includes a brief introduction to cometary science, and describes how amateur and professional astronomers are dealing with Halley's Comet. A special feature of this book is the set of four microcomputer programs for the BBC model B and Spectrum ZX81 (which are also available separately on cassettes), showing, for example, the changing position of Halley's Comet and the Earth relative to the Sun at different dates during the comet's passage.

*736 Tullius, John
THE *SCIENCE DIGEST* BOOK OF HALLEY'S COMET
New York: Avon, 1985, ISBN 0-380-89527-7 pbk. 136p.

The cover of this book promises 'All the Latest Scientific Information: complete chart showing when and where to get the best view of the comet . . .', but the chart is very small and the path of Halley's Comet from November 1985 to May 1986 is out of focus. (From *S&T* Aug. 1985, p 131).

Late entries (not seen)

Dickinson, Terence
Halley's Comet: a mysterious visitor from outer space (Barrington, NJ: Edmund Scientific, 1985, 125p).

Etta, Roberta and Stuart Schneider
Halley's Comet: memories of 1910 (New York: Abbeville Press, 1985, ISBN 0-89659-588-9, 96p).

Greiner, Keith
Halley's Comet: a viewer's essential guide and notebook (Des Moines, IA: Better Life Products, 1984, ISBN 0-9614258-0-6 pbk; 0-9614258-1-4 (European ed.) pbk. 33p).

Gropman, Donald and Kenneth Mervis
Comet fever (New York: Simon & Schuster (Fireside Books), 1985, ISBN 0-671-60307-8).

Halley's Comet star charts book (Northbrook, IL: Hubbard Scientific, 1985, ISBN 0-8331-0433-0 pbk).

Hart, Matthew
A viewer's guide to Halley's Comet (New York: Pocket Books, 1985, ISBN 0-671-49841-X pbk).

Meyer, Ben
Halley's Comet finder (New York: Putnam, 1985, ISBN 0-399-51137-7 pbk, 80p).

NASA Goddard Space Flight Center
Comet Halley returns: a teacher's guide 1985–1986 (Washington, DC: NASA, 1985[?], NASA EP-197 pbk, 41p).

The New York Times *guide to the return of Halley's Comet* (New York: Times Books, 1985, ISBN 0-8129-1148-2; 0-8129-6347-4 pbk, 244p).

Ottewell, Guy and Fred Schaaf
Mankind's comet (Greenville, SC: Furman University Physics Dept., 1985, 200p, Q).

Photographing Halley's Comet (Havant: Space Frontiers, 1985, pbk. 18p).

7 Astronautics

Reference Works

737 Angelo, Joseph A., jr.
THE DICTIONARY OF SPACE TECHNOLOGY
New York: Facts on File, 1982, ISBN 0-87196-583-6; London: Frederick Muller, 1982, ISBN 0-584-95011-X. 383p; illus, diags.

Contains over 1,500 entries covering not only astronautics, but also many complementary terms from physics, astronomy, chemistry, and other physical sciences and engineering disciplines. The length of the entries ranges from very brief definitions to articles of several pages (e.g. Spacelab, 15 pages). Includes a guide to potential future US and other space programmes, but information from the Soviet Union is limited.

738 Furniss, Tim
MANNED SPACEFLIGHT LOG
London: Jane's, 1983, ISBN 0-86720-631-3 pbk. 160p; illus. Index (1). Q

A chronological presentation of the first hundred or so US and Soviet manned space missions, 1960–82. Each entry gives basic mission data and a brief descriptive account. Profusely illustrated, but includes many publicity photographs of astronauts, which present a rather artificial appearance. The index lists only the missions, not the astronauts and cosmonauts.

739 Furniss, Tim
SPACE FLIGHT: THE RECORDS
Enfield: Guinness Superlatives, 1985, ISBN 0-85112-451-8 pbk. 168 + 8p; illus (some col.). Glos (1), index (2).

This comprehensive compendium of manned space flight data is intended to fill the gap of factual information in news coverage of space activities. Five parts detail all manned flights, space firsts, machines, space travellers (career and personal information), and provide tables of other data.

740 Gatland, Kenneth (consultant and principal author)
THE ILLUSTRATED ENCYCLOPEDIA OF SPACE TECHNOLOGY: A COMPREHENSIVE HISTORY OF SPACE EXPLORATION
London: Salamander Books, 1981, ISBN 0-86101-075-2; New York: Crown (Harmony Books), 1981, ISBN 0-517-54258-7. 289p; illus & diags (mostly col.). Bibl (x), index (2). Q

This collection of 21 articles by writers well known in the field, provides an excellent introduction for the general reader, and a useful reference for the enthusiast. Coverage is international. The book is well illustrated, with many colour photographs and explanatory diagrams, and includes two large pull-out sections featuring space launchers and space stations. The 'space diary' is a chronology of spaceflight from 360 BC to the present! (A revised impression, 1984, adds 12 pages to the space diary).

741 Lewis, Richard S. (principal author)
AN ILLUSTRATED ENCYCLOPEDIA OF SPACE EXPLORATION (UK)/THE ILLUSTRATED ENCYCLOPEDIA OF THE UNIVERSE: UNDERSTANDING AND EXPLORING THE COSMOS (US)
London: Salamander Books, 1983, ISBN 0-86101-181-3; New York: Crown (Harmony Books), 1983, ISBN 0-517-55109-8. 320p; illus & diags (mostly col.). Glos (2), bibl (x), index (3). Q

An attractively illustrated companion volume to (740), which covers space science and the wider implications of space technology. The 19 chapters are divided into three sections: 1. extra-solar astronomy and the Sun, 2. the Solar System (both sections emphasizing the technical developments which have led to the astronomical discoveries), 3. future activity, including the exploitation of the resources of the Solar System, space industries, and further exploration. Ends with a ten-page diary of space discovery.

742 King-Hele, D.G., J.A. Pilkington, H. Hiller, D.M.C. Walker and A.N. Winterbottom
THE RAE TABLE OF EARTH SATELLITES, 1957–1982 (2nd ed.)
London: Macmillan, 1983, ISBN 0-333-35374-9; New York: Wiley, 1983, ISBN 0-471-87909-6. 753p. Index (3). Q

A chronological list of the 2,389 launches of satellites and space vehicles between 1957 and the end of 1982 (supplemented by an alphabetical index), based on the records of the Royal Aircraft Establishment, Farnborough. The book also notes fragments associated with a launch and space vehicles that escaped from Earth's influence, to give a total of over 13,000 entries. The introduction explains the development of the table, the number and purpose of launches, nomenclature, orbits, and provides a guide to the table, and describes how it was compiled. The information in the table includes name and designation, launch date, lifetime, descent date, shape and weight, and orbital data.

743 TRW SPACE LOG
Redondo Beach, CA: TRW Electronics & Defense, 1960–, annual, pbk. About 125p; col. illus.

An unofficial record of space launches from 1957 to date, presented chronologically in tabular form. An introductory section gives brief details of the past year's spacecraft. Compiled from data released by NASA, the Department of Defense, the Library of Congress, the United Nations, the Royal Aircraft Establishment and other official sources. The summary listing gives name, international designation, originator, launch data, weight, orbital data, and current status. Includes launches which failed to orbit. (Write to: The editor, Space Log, TRW Electronics & Defense, One Space Park, Mail Station E2/5062, Redondo Beach, CA 90278, USA).

744 Turnill, Reginald
JANE'S SPACEFLIGHT DIRECTORY
London: Jane's, 1984, ISBN 0-7106-0208-1. 311p; illus, diags. Index (3). Q

This recent addition to Jane's invaluable series of reference works is a comprehensive guide to world space activities past and present, with notes on future programmes. The introduction provides a survey of recent developments and trends. Covers space programmes by nation, with separate treatment of international ventures, military space, launchers, the Solar System (exploration records and mission proposals), world space centres, spacemen (and women) — flight records and brief biographical sketches, space contractors. This huge compilation of historical information, mission records, costs, results, and likely future developments is enhanced by numerous illustrations.

745 Turnill, Reginald
THE OBSERVER'S SPACEFLIGHT DIRECTORY
London: Frederick Warne, 1978, ISBN 0-7232-2051-4. 384 + 8p; illus (some col.), diags. Index (2).

Contains details of space programmes in a listing by country or international organization, giving details of many individual missions. Includes tables of major manned and unmanned flights, and separate listings of astronauts and cosmonauts. (This book expands and updates the information in *The Observer's book of manned spaceflight* and *The Observer's book of unmanned spaceflight*).

746 Wells, Helen T., Susan H. Whiteley and Carrie E. Karegeannes
ORIGINS OF NASA NAMES
Washington, DC: NASA, 1976, NASA SP-4402 pbk. 227p; illus. Bibl (3), index (3).

This compilation is limited to the names of approved projects through 1974; it does not describe the projects or give technical details. The entries are arranged in six alphabetically ordered sections: launch vehicles, satellites, space probes, manned space flight, sounding rockets, and NASA installations.

747 Wilding-White, T.M.
JANE'S POCKET BOOK [OF] SPACE EXPLORATION
London: Macdonald & Jane, 1976, ISBN 0-356-08406-X; New York: Collier Macmillan, 1977, ISBN 0-02-080660-4. 238p; illus, diags. Index (1).

A descriptive catalogue of spacecraft, divided into four sections: manned, interplanetary, scientific and applications, and launch vehicles.

General

748 Arnold, H.J.P.
IMAGES FROM SPACE: THE CAMERA IN ORBIT
Oxford: Phaidon, 1979, ISBN 0-7148-2017-2 pbk. 70p; illus (mostly col.). Q

An extra-large format picture book which attempts 'to outline the essence of spaceflight in visual terms'. Six preliminary pages contain notes and captions for the 64 full-page illustrations, which include some astronomical images and some artists' impressions of future possibilities.

749 Bluth, B.J. and S.R. McNeal (ed.)
UPDATE ON SPACE, VOL. 1
Granada Hills, CA: National Behaviour Systems, 1981, ISBN 0-937654-00-0 pbk. 196p; illus, diags.

A collection of eight articles by people involved in the US

space programme and related activities, intended to provide information for the general public to keep them aware of developments, and to provide the necessary knowledge for decision making. Covers topics such as space stations, the benefits of industry in space, and the military use of space.

*750 Cheston, T. Stephen and David C. Webb
SPACE HUMANIZATION SERIES
Washington, DC: Institute for the Social Science of Space, 1979, ISBN 0-934144-01-X. 125p.

Begins a series that represents a concentrated effort by the Institute to 'stimulate social science and humanities scholarship and reflection on space-related topics'. Coverage includes space industrialization, public policy, privacy in space as a conflict of interest, and space agriculture. (From Blashfield (15), p 51).

751 Chipman, Ralph
THE WORLD IN SPACE: A SURVEY OF SPACE ACTIVITIES AND ISSUES
Englewood Cliffs, NJ: Prentice-Hall, 1982, ISBN 0-13-967745-3. 689p; illus, diags. Index (3).

Contains material based on a series of 'Background Papers' prepared for the Second United Nations Conference on the Exploration and Peaceful Uses of Outer Space, or UNISPACE 82, held 9–12 August 1982 in Vienna. It is a review and assessment of developments in space science and technology, not for the casual reader, but a useful compendium for the enthusiast, which brings together many different viewpoints.

752 Cooke, Hereward Lester, (with James D. Dean)
EYEWITNESS TO SPACE
New York: Abrams, 1971, ISBN 0-8109-0112-9. 227p; illus (mostly col.). Q

A large-scale collection of space art, which presents 258 works by 47 artists, representing the fruits of the programme begun by NASA in 1963 to supplement the photographic record in capturing the spirit of the manned space programme.

753 Couper, Heather
EXPLORING SPACE
London: Sundial Books (St. Michael), 1980, ISBN 0-906320-45-3. 96p; col. illus. Index (1). Q

A profusely illustrated guide for readers of all ages, written in a racy style at a fairly basic level.

754 Deudney, Daniel
SPACE: THE HIGH FRONTIER IN PERSPECTIVE
Washington, DC: Worldwatch Institute, 1982, ISBN 0-916468-49-6 pbk. 72p. Bibl (2).

One of a series of booklets produced by 'an independent non-profit research organization created to analyse and to focus attention on global problems'. These Worldwatch Papers (of which this is number 50) are intended for 'a worldwide audience of decision makers, scholars and the general public'. A readable account of our involvement with space, which looks at the potential benefits, but also points out the dangers of human folly.

755 Dickson, Paul
OUT OF THIS WORLD: AMERICAN SPACE PHOTOGRAPHY
New York: Delacorte, 1977, ISBN 0-440-06568-2; New York: Dell (Delta), 1977, ISBN 0-440-56756-4 pbk. 158p; illus (some col.). Q

The first part of this book (about 60 pages) covers the development of US space photography, attempting to emphasize the importance of this aspect of space exploration (*see also* 829). The second part presents the author's personal selection of images, covering a wide range of astronomical and astronautical topics, including a number of images of Earth from space.

756 EUROPE INTO SPACE
Paris: European Space Agency, 1983, pbk. 48p; col. illus & diags. Q

An introduction to European space activities for the general reader, mostly suitable also for younger readers. Provides a guide to ESA, its projects (such as the International Ultraviolet Explorer (IUE), Exosat, Giotto, Space Telescope participation), Earth observation, communication satellites, the Ariane launcher, Spacelab, and some supporting activities. Includes notes on future projects. (*See also* 892).

Late entries: *Forward to the future* (Paris: ESA, [1985], pbk, 51p, Q), an updated version of the above publication. Peter Haining, *Eyewitness to the galaxy: Britain's contribution to research in space* (London: W.H. Allen, 1985, ISBN 0-491-03610-8, 222p), a brief, colourful account.

757 Freeman, Michael
SPACE TRAVELLER'S HANDBOOK: EVERY MAN'S COMPREHENSIVE MANUAL TO SPACE FLIGHT
London: Hamlyn, 1979, ISBN 0-600-38419-5; New York: Sovereign Books, 1979, ISBN 0-671-96147-0. 208p; illus & diags (mostly col.).

Glos (1), index (2). Q

Basically a picture book, but presented as a guide for potential space travellers as though written in the year 2061. The book's four sections cover mission preparation, hardware and spacecraft, control and guidance, and the 'geography of space'. A colourful mixture of fact and speculation.

758 Geis, Larry and Fabrice Florin, (ed. with Peter Beren and Aidan Kelly)
WORLDS BEYOND: THE EVERLASTING FRONTIER (pbk)/ MOVING INTO SPACE: THE MYTHS AND REALITIES OF EXTRATERRESTRIAL SPACE
Berkeley, CA: And/Or Press, 1978, ISBN 0-915904-36-5 pbk; New York: Harper & Row, 1980, ISBN 0-06-080499-8. 301p; illus, diags.

A compendium of fact and speculation about various aspects of space: the future of humans in space, industrialization, extraterrestrial life, UFOs, and 'space age myths' (mind, spirit and intelligence in space).

759 Grey, Jerry
ENTERPRISE
New York: Morrow, 1979, ISBN 0-688-03462-4. 288p; illus. Index (2).

A history of, and future prospects for the US space programme, by a NASA staff member, aimed at the general reader. Appendices include a chronology and a list of acronyms.

760 Hartz, Theodore R. and Irvine Paghis
SPACEBOUND
Ottawa: Canadian Government Publishing Centre, 1982, ISBN 0-660-11187-X pbk. 188p; illus & diags (mostly col.). Bibl (1), glos (2).

An official account, in a popular format, of the Canadian space programme. Describes the five different ventures which involved Canadian-built satellites, at first concentrating on studies of the atmosphere, but later turning to communications.

761 Ince, Martin
SPACE
London: Sphere, 1981, ISBN 0-7221-4910-7 pbk. 215 + 8p; illus. Bibl (2).

A general survey of space activities for the lay reader. Includes a good descriptive bibliography, but no index.

762 [Mason, Robert Grant (ed.)]
LIFE IN SPACE
Alexandria, VA: Time-Life, 1983, ISBN 0-8094-4900-5. 304p; illus (mostly col.). Bibl (1), index (2). Q

A huge, colourful celebration of mankind's accomplishments in space, the book contains countless photographs and only minimal text. Includes many unusual pictures, showing the human side of space flight as well as the technology.

763 Moché, Dinah L.
LIFE IN SPACE
London: Hamlyn, 1979, ISBN 0-600-35571-3. 160p; illus (mostly col.). Index (1). Q

A profusely illustrated guide to recent developments in space exploration for readers of all ages. Includes artwork and movie stills, as well as photographs.

*764 Ward, R.J., jr.
THE LIGHT STUFF: SPACE HUMOR — FROM SPUTNIK TO SHUTTLE
[?]: Jester Books, 1982, pbk. 295p.

A book of anecdotes and stories concerning the space business, highlighting much of the repartee which accompanied the American space programme, and providing much insight into the attitudes of those involved. (From *Spaceflight* Dec. 1982, p 474).

Astronautics plus . . .

*765 Bainbridge, William Sims
THE SPACEFLIGHT REVOLUTION: A SOCIOLOGICAL STUDY (2nd ed.)
Melbourne, FL: Krieger, 1983, ISBN 0-89874-501-2. 249p.

A study of the social movement which brought about the spaceflight revolution, including accounts of such organizations as the British Interplanetary Society, and a discussion of the 'science fiction subculture'. Though successful the movement had, from the first, aimed at the attainment of ever-expanding manned space flight. (From *Space Education* Oct. 1983, p 288).

*766 Bell, Trudy E.
UPWARD: STATUS REPORT AND DIRECTORY OF THE AMERICAN SPACE INTEREST MOVEMENTS, 1984–85
New York: T.E. Bell, 1985, pbk. 105p.

The third in a biennial series of surveys assesses the aims and effectiveness of American space-interest organizations. An appendix tabulates detailed information on 48 groups. (From *S&T* Aug. 1985, p 133).

767 Berman, Arthur I.
SPACE FLIGHT
Garden City, NY: Doubleday (Anchor Press), 1979, ISBN 0-385-02754-0 pbk. 206 + 16p; illus, diags. Index (2).

A popular account of the physics involved in space flight, which begins with a discussion of escape velocity and orbital flight, then goes on to the use of the Moon as a site for low-gravity launches. Also considers 'solar rails', relativity and gravitation, and possible methods of achieving interstellar flight.

768 Elliot, Jeffrey M.
THE FUTURE OF THE SPACE PROGRAM: LARGE CORPORATIONS AND SOCIETY
San Bernadino, CA: Borgo Press, 1981, ISBN 0-89370-140-8; 0-89370-240-4 pbk. 64p.

Discussions with 22 science fiction writers on the role that they can and should play in working with America's major corporations to plan for the future in space.

769 Emme, Eugene M. (ed.)
SCIENCE FICTION AND SPACE FUTURES PAST AND PRESENT
San Diego, CA: Univelt, 1982, ISBN 0-87703-172-X; 0-87703-173-8 pbk. 270p; illus, diags. Bibl (3), index (2).

The proceedings of the Third AAS History Symposium, held in March 1981 in Pentagon City, Virginia. In the first part of the book five papers examine the relationship between SF and astronautics in its broadest sense, including SF in the cinema, space art, and "The impact of SF on attitudes towards technology". The second part is a panel discussion on space science futures, while the third considers religious and ethical implications of our involvement with space, and includes a 33-page 'eclectic bibliography on the history of space futures' by the editor.

770 Harris, Gordon L.
SELLING UNCLE SAM
Hicksville, NY: Exposition, 1976, ISBN 0-682-48481-4. 222 + 8p; illus. Index (2).

An insider's anecdotal account of public relations mostly associated with the US space programme, covering the author's work at Huntsville, Alabama; Washington, DC and Cape Canaveral.

*771 Malz, Leo
SPACE STAMPS
Johnstown, PA: American Topical Association, 1985. 86p.

This handbook lists all known space stamps and souvenir sheets, some 6,000 or so, issued by all countries of the world and arranged alphabetically. Includes selected astronomy issues and contains a section by Lee Saegesser describing spacecraft, missions, programmes and personalities. (From *Spaceflight* June 1985, p 287, and *Space Age Times* March/April 1985, p 25–6).

An earlier guide is, R.F. Kitchen, *First men on the Moon: handbook of stamps* (New York: Vantage, 1976, ISBN 0-533-01957-5, 139p).

772 NASA
SPINOFF
Washington, DC: NASA, annual, pbk. Illus (mostly col.). Q

This publication describes 1. NASA's current major projects, their dividends and indirect benefits to the public, 2. a selection of spinoff products and processes, and the NASA technology from which they are derived, and 3. mechanisms employed to encourage and facilitate the application of new technologies developed in the course of NASA research.

773 Parkinson, Bob and R.A. Smith
HIGH ROAD TO THE MOON: FROM IMAGINATION TO REALITY
London: British Interplanetary Society, 1979, pbk. 120p; illus. Q

This book describes the development and realization of the idea of travel to the Moon, using the imaginative paintings of R.A. Smith, who died in 1959. The second part of the book moves beyond the Moon to consider possible future developments in space travel, using the artist's more speculative pictures.

774 Sagan, Carl, F.D. Drake, Ann Druyan, Timothy Ferris, Jon Lomberg and Linda Salzman Sagan
MURMURS OF EARTH: THE VOYAGER INTERSTELLAR RECORD
New York: Random House, 1978, ISBN 0-394-31047-5; 0-394-28396-1 pbk; London: Hodder & Stoughton, 1978, ISBN 0-340-24423-2. 275 + 8p; illus (some col.). Bibl (x), index (2).

The story behind the record despatched with the Voyager

spacecraft: gold-coated phonograph records containing a selection of items, including photographs encoded in audio form, music, and greetings in various languages, intended to represent Earth and its culture. Includes a complete catalogue of the content of the records.

775 Smith, Delbert D.
SPACE STATIONS: INTERNATIONAL LAW AND POLICY
Boulder, CO: Westview, 1979, ISBN 0-89158-654-7. 264p. Bibl (3), index (2).

A detailed, though largely non-technical discussion of the legal issues relating to space stations, including the problem of defining what is a space station.

*776 Sheffield, Charles and Carol Rosin
SPACE CAREERS
New York: Morrow, 1984, ISBN 0-688-03182-X; 0-688-03256-7 pbk. 240p; illus. Bibl, index.

Vocational guidance.

*777 Weber, Ronald
SEEING EARTH: LITERARY RESPONSES TO SPACE EXPLORATION
Athens, OH: Ohio University Press, 1985, ISBN 0-8214-0791-0. 138p. Bibl (3), index.

This book investigates various writers' attitudes toward, and perception of space travel. Featured are excerpts from Tom Wolfe, Carl Sagan, James Michener, Norman Mailer, Ken Kesey, Ray Bradbury and Ben Bova. (From *Astronomy* July 1985, p 57).

Periodicals

778 L5 NEWS
(L5 Society, 1060 E. Elm, Tucson, AZ 85719, USA), 1976–, monthly, ISSN 8756-5331. 20p; illus (col. cover), diags. Q

L5 News is available to members of the L5 Society, an organization founded in 1975 to promote the development of the 'nearly limitless resources of the Solar System to permit the establishment of a space based civilization [which is] economically self sufficient and supported by a free enterprise structure'. The magazine contains articles and features which reflect this viewpoint. Includes book reviews and correspondence. 'L5' refers to the fifth of five stable orbital locations in the Earth-Moon system where the first settlement might be constructed (*see* 959).

779 NASA ACTIVITIES
(Superintendent of Documents, US Government Printing Office, Washington, DC 20402, USA), 1970–, monthly. 20p; illus.

Contains brief news items about NASA projects and personnel.

780 SPACE AGE TIMES
(International Membership Committee, United States Space Education Association, International Headquarters, 746 Turnpike Road, Elizabethtown, PA 17022-1161, USA), 1974–, bi-monthly, ISSN 0741-4587. About 30p; illus, diags. Ads. Q

'The USSEA is a worldwide non-profit, non-partisan organization dedicated to promoting peaceful space exploration, seeking to achieve goals through both professionals and the layman'. The magazine contains articles and news covering all aspects of space exploration, including space benefits and news of space interest groups. Includes book reviews and correspondence. The USSEA also produces *Update: the bulletin of important space news, politics and information* (1980–), a two-page monthly newsletter.

781 SPACE EDUCATION
(British Interplanetary Society, 27/29 South Lambeth Road, London, SW8 1SZ, UK), 1981–, 2 issues/year, ISSN 0261-1813. Illus, diags. Q

A magazine 'devoted to the role of education in Space Exploration and related subject areas, being particularly concerned with the potentialities of space developments as they offer new career prospects and increasingly influence human affairs and destiny'. Aims to assist teachers, and provide career information and general background material for students just below and at university level. Includes correspondence, book and media reviews.

782 SPACEFLIGHT
(British Interplanetary Society, 27/29 South Lambeth Road, London, SW8 1SZ, UK), 1959–, 10 issues/year, ISSN 0038-6340. Illus, diags. Ads. Q

The BIS produces three periodicals: the *Journal* is devoted largely to technical articles for specialists, including historical studies; *Space Education* (781); and *Spaceflight*, which includes items of a more popular nature, but is intended for those with a serious interest in space. It covers all aspects of astronautics, including regular news features from the US and elsewhere, a continuing chronology ("Milestones"), a regular listing of satellite and spacecraft launches ("Satellite digest"), feature articles, correspondence and book notices. (*See also* 773 and 832).

The BIS was responsible for an engineering study of a possible unmanned interstellar spacecraft, the results of which were published as *Project Daedalus* (BIS, 1978).

783 SPACEFLIGHT NEWS
(1 Wothorpe Road, Stamford, Lincolnshire, PE9 2JR, UK), 1985–, bi-monthly, ISSN 0268-4713. 48p; illus (some col.), diags. Q

A colourful, well illustrated magazine for the space enthusiast, providing news of space activities round the world, but emphasizing the role of the Space Shuttle. Includes feature articles on special topics, history and future prospects; book reviews and correspondence. This magazine succeeds *Shuttle Story*, a six (and one extra)-issue part-work (*see* 906).

784 SPACE FOR ALL PEOPLE
(Progressive Space Forum, 1724 Sacramento Street, Suite 105, San Francisco, CA 94109, USA), 1980–?, 2–3 issues/year. 8p; diags. Q

The organ of the Progressive Space Forum (formerly known as Citizens for Space Demilitarization) contains news and articles promoting the peaceful use of space, while dealing with the wider aspects of space technology and related legal, political and social matters. (This periodical may have been discontinued).

785 SPACE NEWS
(Space News Publishing Co., PO Box 66521, Baton Rouge, LA 70896, USA), monthly +, ISSN 0277-6030. 4p; illus. Ads. Q

This brief newsletter contains reports on astronautics and astronomy, written in a popular style. Published when appropriate, though basically monthly, with additional issues for major events and discoveries.

786 SPACE WORLD
(National Space Institute, Membership Dept., PO Box 7537, Ben Franklin Station, Washington, DC 20044, USA), 10 issues/year, ISSN 0038-6332. About 40p; illus & diags (some col.). Ads. Q

Available to members of the NSI, this magazine for the space enthusiast contains general news, views and features, though slanted toward US matters. Includes letters, listings of events, an annotated list of new publications ("Extraterrestrial resources"). A newsletter, *Insight*, is bound into members' copies.

787 WORLD SPACEFLIGHT NEWS
(Box 98, Sewell, NJ 08080-0098, USA), 1983–, monthly, ISSN 0737-8548. 16p; diags. Q

This periodical concentrates on very detailed coverage of Space Shuttle missions. It contains pre-flight guides, including crew activity plans, comprehensive descriptions of payloads and experiments, and astronaut career profiles; also post-flight reports. Provides coverage of other scientific and commercial spacecraft, mostly US, but includes the Ariane launch schedule, revised monthly. Difficult to read in parts because of the use of huge blocks of very small print.

Practical

788 Eisberg, Robert and Wendell Hyde
COUNTDOWN: SKYDIVER, ROCKET AND SATELLITE MOTION ON PROGRAMMABLE CALCULATORS
Portland, OR: Dilithium Press, 1979, ISBN 0-918398-26-6 pbk. 107p; diags.

How to predict the motion of model rockets, Earth satellites, planets, and alpha particle scattering from atomic nuclei. Uses only the most basic mathematics and physics, and assumes no familiarity with a programmable calculator. An appendix lists the programs described in the text.

789 Garrett, Roger
THE COMPLETE *STAR SHIP*: A SIMULATION PROJECT
Portland, OR: Dilithium Press, 1978, ISBN 0-918398-10-X. 122p; diags. Glos (1).

An introduction to computer simulation (the representation of physical systems) using a starship simulation game as an example. After discussing software design and program structure, the author goes on to describe the program structuring method used for the STAR SHIP project, one which is not associated with any particular programming language. The game is based on *Star Trek*'s *Enterprise*. The final chapter provides hints on the implementation of the program.

790 King-Hele, Desmond
OBSERVING EARTH SATELLITES
London: Macmillan, 1983, ISBN 0-333-33041-2; New York: Van Nostrand, 1983, ISBN 0-442-24877-6. 184p; illus, diags. Bibl (1), index (2).

'The aim of this book is to provide a readable guide to the methods of tracking the 5,000 or so artificial satellites in orbit about the Earth, with particular emphasis on do-it-yourself visual observing'.

About 200 of these satellites are visible to the naked eye in the right conditions, while about 2,000 can be seen with 7 × 50 binoculars. Written in an engaging style, the book covers launches, orbital behaviour, types of satellite, timing, photography, radio tracking, the Earth's form and atmosphere. (This is a completely rewritten version of a book published in 1966).

791 Miles, Howard (ed.)
ARTIFICIAL SATELLITE OBSERVING AND ITS APPLICATIONS
London: Faber, 1974, ISBN 0-571-09600-X; New York: Elsevier, 1974, ISBN 0-444-19544-0. 216 + 8p; illus, diags. Bibl (1), index (2).

A guide for the enthusiast, a little more technical than (790), which provides background information as well as practical guidance, and includes the necessary mathematical information for preparing predictions, and for other applications.

792 Prentiss, Stan
SATELLITE COMMUNICATIONS
Blue Ridge Summit, PA: TAB Books, 1983, ISBN 0-8306-1632-2 pbk. 280p; illus, diags. Index (1).

A guide for electronics hobbyists, which describes systems currently in use and provides instructions for the construction of various Earth-based communication systems. Claims to answer every question that might be asked.

793 Stine, G. Harry
HANDBOOK OF MODEL ROCKETRY (5th ed.)
New York: Arco, 1983, ISBN 0-668-05358-5; 0-668-05360-7 pbk. 367p; illus, diags.

'People will go to the stars someday, and model rocketry may help pave the way', writes the author of the standard work for the enthusiast. Covers tools and workshop techniques, construction, motors, ignition and ignition systems, launchers and launching stability, aerodynamics, multistaged rockets, recovery devices, glide recovery, payloads, scale models, and altitude determination.

History

General

794 Armstrong, Neil A.
CHANGE IN THE SPACE AGE
Edinburgh: Edinburgh University Press, 1973, ISBN 0-85224-239-5. 24p; illus.

A transcript of the Mountbatten Lecture delivered by the first man to set foot on the Moon, at the University of Edinburgh in 1972, which reviews the first decade of the space age.

795 Baker, David
CONQUEST: A HISTORY OF SPACE ACHIEVEMENTS FROM SCIENCE FICTION TO THE SHUTTLE
Leicester: Windward, 1984, ISBN 0-7112-0403-9; Salem, NH: Merrimack (Salem House), 1985, ISBN 0-947703-00-4. 191p; illus (mostly col.). Index (2). Q

A highly illustrated account aimed at the general reader with little or no previous knowledge. Ten tables at the back of the book include details of manned flights, cosmonauts, astronauts, and launch vehicles.

796 Baker, David
THE HISTORY OF MANNED SPACEFLIGHT
London: New Cavendish, 1981, ISBN 0-904568-30-X; New York: Crown, 1982, ISBN 0-517-54377-X. 544p; illus (some col.), diags. Index (2). Q

This massive book is intended for the general reader as a complete history of the first 20 years of manned spaceflight, and for the student as a starting point for specific aspects. The emphasis is on US achievements because Soviet material is less readily available, much of what is available is speculative, and because the wide-ranging US achievements in exploration and discovery are not mirrored by Soviet efforts. Deals with the technology of spaceflight; does not cover the astronomical discoveries made. The author's style is a little long-winded in places. Includes eight pages of space project data, and 29 tables of mission data. A bibliography would have been useful.

797 Baker, David
THE ROCKET: THE HISTORY AND DEVELOPMENT OF ROCKET AND MISSILE TECHNOLOGY
London: New Cavendish, 1978, ISBN 0-904568-17-2; New York:

Crown, 1978, ISBN 0-517-53404-X. 277p; illus (some col.), diags. Glos (1), bibl (1), index (2). Q

A huge, ambitious work, intended for the student of this aspect of the history of technology, as well as the general reader. The eleven chapters are arranged by topic rather than chronologically. Contains a 40-page compendium of technical histories of 98 launch vehicles and ballistic missiles, alphabetically arranged by country. Includes a world rocket and missile inventory: over six pages of tables listing vehicles by type, e.g. research rockets, anti-ballistic missiles.

798 Blaine, J.C.D.
THE END OF AN ERA IN SPACE EXPLORATION: FROM INTERNATIONAL RIVALRY TO INTERNATIONAL COOPERATION
San Diego, CA: American Astronautical Society, 1976, ISBN 0-87703-084-7. 199p; illus, diags. Bibl (2), index (2).

'A non-technical presentation of the events and achievements of the first era of the Space Age characterized by the great advances in technology engendered in part by a contest between the Soviet Union and the United States in the exploration of . . . the Solar System'. The author's aim is to emphasize the need for international co-operation in the exploration and use of space.

799 Durant, Frederick C. and George S. James (ed.)
FIRST STEPS TOWARD SPACE
Washington, DC: Smithsonian Institution, 1974. 307p; illus, diags. Bibl (x), index (2).

Twenty-seven papers based on symposia held in 1967 in Belgrade, and 1968 in New York, organized by the International Academy of Astronautics, covering various aspects of the history of astronautics. Often technical, they include some memoirs by the people personally involved.

800 Hutton, Richard
THE COSMIC CHASE
New York: New American Library (Mentor), 1981, ISBN 0-451-61925-0 pbk. 225p. Bibl (2), index (2).

A history of space travel, which concentrates on the ideas and the politics, rather than the technology or the chronicling of every mission. Begins with the German V-2 project in the second world war. Covers Soviet spaceflight, but has a US bias.

801 Needell, Alan A. (ed.)
THE FIRST 25 YEARS IN SPACE
Washington, DC: Smithsonian Institution Press, 1983, ISBN 0-87474-668-X. 152p; illus. Bibl (x), index (2).

The proceedings of a symposium held in October 1982 at the National Academy of Sciences in Washington. In three sections, with introductions, comments and discussions, the book provides a non-technical look at: 1. the motivations for space activity, 2. the practical dimensions of space, and 3. science and space.

802 Nicolson, Iain
SPUTNIK TO SPACE SHUTTLE: 25 YEARS OF THE SPACE AGE
London: Sidgwick & Jackson, 1982, ISBN 0-283-98891-6; New York: Dodd, Mead, 1983, ISBN 0-396-08231-9. 207p; illus, diags. Bibl (1), index (2).

A history of space flight for the general reader, which provides background information on astronomy and physics, and deals with methods of propulsion. The main part of the book covers unmanned exploration, astronomical advances made possible by observations from outside the Earth's atmosphere and beyond the visible part of the electromagnetic spectrum, the uses of satellites, man in space, space stations, the Space Shuttle, and future possibilities. An appendix provides a chronology of the highlights of space activity 1957–82.

803 Osman, Tony
SPACE HISTORY
London: Michael Joseph, 1983, ISBN 0-7181-2190-2; New York: St. Martin's, 1983, ISBN 0-312-74945-7. 216 + 8p; illus (some col.), diags. Glos (1), bibl (1), index (2).

The story of space flight, for the general reader, presented chronologically from early speculation, through the reality, to future possibilities.

804 Schauer, William H.
THE POLITICS OF SPACE: A COMPARISON OF THE SOVIET AND AMERICAN SPACE PROGRAMS
New York and London: Holmes & Meier, 1976, ISBN 0-8419-0185-6. 317p; Bibl (3), index (2).

A scholarly study of the organization and funding of the space efforts of the two major world powers, devoted mostly to the period after 1957.

805 Trux, John
THE SPACE RACE: FROM SPUTNIK TO SPACE SHUTTLE: THE STORY OF THE BATTLE FOR THE HEAVENS
Sevenoaks: New English Library, 1985, ISBN 0-450-06046-2. 160p; illus (some col.). Index (1).

A brief, highly illustrated account of Soviet-American rivalry in space.

*806 Vaning, Walker S.
SPACE: 20 YEARS AND COUNTING
Temple, AZ: Payload Technology Inc., 1977, pbk. 70p; illus. Q

807 von Braun, Wernher and Frederick I. Ordway, III
HISTORY OF ROCKETRY AND SPACE TRAVEL (3rd ed.)
New York: Crowell, 1975, ISBN 0-690-00588-1. 308 + 40p; illus (some col.), diags. Bibl (3), index (2). Q

A highly illustrated, authoritatively researched account for the general reader, which considers the parallel tracks of the development of rocketry and space flight, the former providing the impetus for the latter. International in scope, the book covers military as well as scientific applications.

By the same authors is, *The rockets' red glare: a history of rocketry through the ages* (Garden City, NY: Doubleday (Anchor Press), 1976, ISBN 0-385-07847-1).

808 Wilson, Andrew
FRONTIERS OF SPACE
London: Hamlyn (Deans International), 1985, ISBN 0-603-03746-1. 192p; illus (some col.), diags. Index (2). Q

A highly illustrated book aimed at the casual reader, but whose text provides a good, concise historical survey of space flight from its beginnings to the present, with later chapters covering the Space Shuttle, space stations, the military use of space, and future possibilities.

United States

809 Anderson, Frank W.
ORDERS OF MAGNITUDE: A HISTORY OF NACA AND NASA 1915–1980
Washington, DC: NASA, 1981, NASA SP-4403 pbk. 106p; illus.

This concise account for the general reader concentrates on NASA, giving only brief coverage of its predecessor, the NACA (National Advisory Committee for Aeronautics).

Late entry: *NASA 1958–1983: remembered images* (Washington, DC: NASA, 1985[?], NASA EP-200 pbk, 136p, Q), a colourful, pictorial retrospective of NASA's first 25 years.

810 Benson, Charles D. and William Barnaby Faherty
MOONPORT: A HISTORY OF APOLLO LAUNCH FACILITIES AND OPERATIONS
Washington, DC: NASA, 1978, NASA SP-4204 pbk. 636p; illus (1 col.), diags. Bibl (3), index (3).

This detailed account attempts to untangle the complexities of this story from the beginnings of the design up to and including the final launch.

811 Bilstein, Roger E.
STAGES TO SATURN: A TECHNOLOGICAL HISTORY OF THE APOLLO/SATURN LAUNCH VEHICLES
Washington, DC: NASA, 1980, NASA SP-4206 pbk. 511p; illus (1 col.), diags. Bibl (3), index (3).

The story of the development of the rocket which served the successful Apollo project. Deals with the technical and managerial aspects of the endeavour, not the political side.

812 Brooks, Courtney G., James M. Grimwood and Loyd S. Swenson, jr.
CHARIOTS FOR APOLLO: A HISTORY OF MANNED LUNAR SPACECRAFT
Washington, DC: NASA, 1979, NASA SP-4205 pbk. 538p; illus, diags. Bibl (3), index (3).

This detailed account concentrates on the planning and construction of the spacecraft, but also provides some coverage of the missions flown.

813 Compton, W. David and Charles D. Benson
LIVING AND WORKING IN SPACE: A HISTORY OF SKYLAB
Washington, DC: NASA, 1983, NASA SP-4208 pbk. 449p; illus (some col.), diags. Bibl (3), index (2).

A detailed account in three parts: 1. concept to decision 1962–69, 2. developments and preparations to fly 1969–73, and 3. the missions and results. Coverage is complete to the end of the manned missions, but not for the results. An appendix provides brief biographies of the Skylab astronauts.

814 Cooper, Henry S.F.
THIRTEEN: THE FLIGHT THAT FAILED/MOONWRECK: 13, THE FLIGHT THAT FAILED (UK pbk)
New York: Dial, 1973, ISBN 0-8037-8765-0; London: Angus & Robertson, 1973, ISBN 0-207-95485-2; St. Albans: Panther, 1975, ISBN 0-586-04165-6 pbk. 199p.

The story of the ill-fated Apollo 13 mission, told in a journalistic style. Following the explosion of an oxygen tank on the way to the Moon, the astronauts survived by using the Lunar Module as a 'lifeboat'.

815 Cortright, Edgar M. (ed.)
APOLLO EXPEDITIONS TO THE MOON
Washington, DC: NASA, 1975, NASA SP-350. 313p; illus & diags (mostly col.). Index (2). Q

This highly illustrated account for the general reader is not restricted to coverage of the lunar missions; each of the contributors, all of whom were involved with the project in some way, recount their part in the Apollo story. Includes a chronology, 1958–72.

816 DeWard, E. John and Nancy
HISTORY OF NASA: AMERICA'S VOYAGE TO THE STARS
Greenwich, CT: Bison Books, 1984, ISBN 0-86124-182-7. 192p; illus, diags (mostly col.). Index (2).

This is a picture book with very brief text, not an official history.

817 Durant, Frederick C., III (ed.)
BETWEEN SPUTNIK AND THE SHUTTLE: NEW PERSPECTIVES ON AMERICAN ASTRONAUTICS
San Diego, CA: Univelt, 1981, ISBN 0-87703-145-2; 0-87703-149-5 pbk. 334p; illus, diags. Bibl (x).

Based on two symposia sponsored by the History Committee of the American Astronautical Society in 1979 and 1980, this volume includes eight essays discussing various aspects of the history of American astronautics, including "Presidents and space", space transportation, and astronautics and art.

818 Ezell, Edward Clinton and Linda Neuman Ezell
ON MARS: EXPLORATION OF THE RED PLANET 1958–1978
Washington, DC: NASA, 1984, NASA SP-4212 pbk. 535p; illus (1 col.), diags. Bibl (3), index (3).

This detailed account covers unmanned missions and early attempts to search for life on Mars, but concentrates on the Viking project.

819 Farmer, Gene and Dora Jean Hamblin
FIRST ON THE MOON: A VOYAGE WITH NEIL ARMSTRONG, MICHAEL COLLINS, EDWIN E. ALDRIN, JR.
Boston: Little, Brown, 1970; London: Michael Joseph, 1970, ISBN 0-7181-0786-3. 434p; illus.

A dramatic account of the first manned Moon landing, concentrating on the personalities and life stories of the astronauts, told largely through reported interviews. Includes a 49-page epilogue, "Beyond Apollo" by Arthur C. Clarke.

820 Goldstein, Laurence (ed.)
THE MOON LANDING AND ITS AFTERMATH
Ann Arbor, MI: *Michigan Quarterly Review* 18 (2), Spring 1979. 363p; illus.

This special issue of a literary magazine explores the impact of the Moon landing of 1969 on the American people, in poetry, art, fiction and essays.

821 Green, Constance McLaughlin and Milton Lomask
VANGUARD: A HISTORY
Washington, DC: NASA, 1970, NASA SP-4202 pbk; Washington, DC: Smithsonian Institution, 1971, ISBN 0-87474-112-2. 309p; illus, diags. Bibl (3), index (3).

A detailed account of the origin, course of development and results of the first American scientific satellite project. Discusses the organizational and political background, notably the competition from the Soviet Union. Intended for the lay reader, but not for casual reading.

822 Grissom, Virgil
GEMINI: A PERSONAL ACCOUNT OF MAN'S VENTURE INTO SPACE
New York: Macmillan, 1968. 212p; illus. Glos (3), index (2).

The story of the Gemini project as told by one of the astronauts most closely involved. (*See also* 823 and 857).

823 Hacker, Barton C. and James M. Grimwood
ON THE SHOULDERS OF TITANS: A HISTORY OF PROJECT GEMINI
Washington, DC: NASA, 1977, NASA SP-4203 pbk. 625p; illus (some col.), diags. Bibl (3), glos (2), index (3).

A detailed account of project Gemini which represented the second stage (after project Mercury, *see* 831) of the vehicle development programme which led to the first lunar landing. Ten manned missions were flown between March 1965 and November 1966.

824 Hall, Cargill
LUNAR IMPACT: A HISTORY OF PROJECT RANGER
Washington, DC: NASA, 1977, NASA SP-4210 pbk. 450p; illus, diags. Bibl (3), index (2).
This detailed account for both specialists and the general reader covers the successes and failures of the NASA unmanned lunar exploration programme.

825 Hallion, Richard P. and Tom D. Crouch (ed.)
APOLLO: TEN YEARS SINCE TRANQUILITY BASE
Washington, DC: Smithsonian Institution Press, 1979, ISBN 0-87474-506-3; 0-87474-505-5 pbk. 174p; illus. Bibl (2). Q
A collection of articles published to celebrate the tenth anniversary of the Apollo 11 Moon landing. Intending to be more than merely self-congratulatory, the writers attempt to correct factual errors, 'dubious interpretation, or subtle romantic embellishment' in the general perception of the event. The 18 pieces survey the Apollo project, its history, implications, achievements and technical aspects. Includes a chronology, mission data, and 'a selective bibliography and research guide', but unfortunately lacks an index.

826 Koppes, Clayton R.
JPL AND THE AMERICAN SPACE PROGRAM
New Haven, CT and London: Yale University Press, 1982, ISBN 0-300-02408-8. 299 + 2p; illus (some col.), diags. Indexes (2).
The history of rocket and space technology at the Jet Propulsion Laboratory in Pasadena, California, from 1936, when it began as the Guggenheim Aeronautical Laboratory, California Institute of Technology (GALCIT). Not an official history, and not primarily a history of science and technology; the author addresses the societal and political issues. Includes coverage of military research.

827 Lewis, Richard S.
THE VOYAGES OF APOLLO: THE EXPLORATION OF THE MOON
New York: New York Times (Quadrangle), 1974, ISBN 0-8129-0477-X. 308 + 16p; illus, diags. Index (2).
This account for the general reader deals with the reasons for going to the Moon, provides background information on the state of lunar

knowledge before Apollo, then moves on to discuss the unmanned missions, and, in more detail, the manned flights, showing what was achieved, and the progressive growth in our knowledge of the Moon.

828 Mailer, Norman
OF A FIRE ON THE MOON (US)/A FIRE ON THE MOON (UK)
Boston: Little, Brown, 1970; London: Weidenfeld & Nicolson, 1970, ISBN 0-297-17952-7; London: Pan, 1971, 0-330-02610-0 pbk; New York: New American Library (Plume), 1982, ISBN 0-452-25377-2 pbk. (No. of pages varies).

A 'literary' writer's view of the Apollo 11 mission. Not a catalogue of technical details, but a story compellingly told in a way that brings the participants to life for the reader.

829 Mirabito, Michael M.
THE EXPLORATION OF OUTER SPACE WITH CAMERAS: A HISTORY OF THE NASA UNMANNED SPACECRAFT MISSIONS
Jefferson, NC: McFarland, 1983, ISBN 0-89950-061-7. 170 + 32p; illus, diags. Bibl (2), index (1).

A scholarly study which 'chronicles the use of television cameras and other visual imaging systems . . . in the exploration of the Solar System's planetary bodies'. Discusses the missions, the visual imaging devices, the technological development of imaging systems and the criteria that influence their design. Covers also image processing and enhancement techniques that converted the transmissions from space into black and white, and colour pictures; the discoveries made through the use of these pictures; potential future planetary missions; and the terrestrial applications of the technology of the space programme.

830 Newell, Homer E.
BEYOND THE ATMOSPHERE: EARLY YEARS OF SPACE SCIENCE
Washington, DC: NASA, 1980, NASA SP-4211 pbk. 497p; illus (1 col.), diags. Bibl (3), index (3).

This detailed history begins with the story of the pioneers of rocketry, but deals mainly with NASA in recounting the technological and managerial development of space science projects and their results, which have greatly enhanced our understanding of the universe.

831 Swenson, Loyd W., James M. Grimwood and Charles C. Alexander
THIS NEW OCEAN: A HISTORY OF PROJECT MERCURY
Washington, DC: NASA, 1966, NASA SP-4201 hardcover & pbk. 681p; illus, diags. Bibl (3), index (3).

A detailed history of America's first manned space project, which became the first stage of the programme which would land the first man on the Moon. (*See also* 823 and 865).

832 Wilson, Andrew
THE EAGLE HAS WINGS: THE STORY OF AMERICAN SPACE EXPLORATION 1945–1975
London: British Interplanetary Society, 1982, pbk. 140p; illus, diags. Q

This account is restricted to manned and interplanetary missions, and allocates more space to certain topics which are less well covered elsewhere; for example, there is less on the Saturn V rocket than on the Jupiter, Redstone and Thor missiles. A little too specialized for the general reader, this book concentrates on the technical aspects of space flight, and on mission planning. (Address: *see* 782).

833 Young, Hugo, Bryan Silcock and Peter Dunn
JOURNEY TO TRANQUILITY: THE HISTORY OF MAN'S ASSAULT ON THE MOON
London: Jonathan Cape, 1969, ISBN 0-224-61784-2; Garden City, NY: Doubleday, 1970. 352 + 32p; illus. Index (2).

This book presents a wider perspective than most of the other books rushed into print after the Apollo 11 Moon landing.

Soviet Union

834 Daniloff, Nicholas
THE KREMLIN AND THE COSMOS
New York: Knopf, 1972, ISBN 0-394-47493-7. 269p; illus. Bibl (2), index (2).

A history of the Soviet space programme, concentrating on the period which began with the launch of Sputnik 1 in 1957.

835 James, Peter N.
SOVIET CONQUEST FROM SPACE
New Rochelle, NY: Arlington House, 1974, ISBN 0-87000-224-4. 256p; illus, diags. Bibl (2), index (3).

The story of the Soviet space programme, based on secret reports prepared by the author, and experience gained during nine years or so as a space systems analyst, and seven years' association with Soviet scientists, engineers, managers and intelligence agents.

836 Johnson, Nicholas L.
HANDBOOK OF SOVIET LUNAR AND PLANETARY EXPLORATION
San Diego, CA: Univelt, 1979, ISBN 0-87703-130-4; 0-87703-131-2 pbk. 262p; illus, diags. Bibl (3), index (2).

A thorough survey of the unmanned Soviet space programme, from the work of Konstantin Tsiolkovsky at the turn of this century up to 1979.

837 Johnson, Nicholas L.
HANDBOOK OF SOVIET MANNED SPACE FLIGHT
San Diego, CA: Univelt, 1980, ISBN 0-87703-115-0; 0-87703-116-9 pbk. 461p; illus, diags. Bibl (x, 3), index (3).

A detailed survey of Soviet manned space flight from Vostok 1, in which Yuri Gagarin became the first human in space, to the long-stay Soyuz-Salyut missions.

838 Oberg, James E.
RED STAR IN ORBIT
New York: Random House, 1981, ISBN 0-394-51429-7; London: Harrap, 1981, ISBN 0-245-53809-7. 272 + 16p; illus. Bibl (2), index (2).

A history of the Soviet space programme for the general reader, based on considerable research and investigation. Presents the stories behind such events as the first man in space, the first woman in space, the Moon race cover-up, secret space cities (elite training villages), and serious setbacks which were not made public at the time. Includes examples of faked photographs, where persons disgraced have been removed. The appendices include biographical sketches of the cosmonauts and other people involved in the story, and a listing of Soviet space shots to March 1978.

839 Popescu, Julian
RUSSIAN SPACE EXPLORATION: THE FIRST 21 YEARS
Henley-on-Thames: Gothard House, 1979, ISBN 0-85638-013-X. 150 + 8p; illus, diags. Glos (2), index (1).

The object of this book 'is to give a systematic and accurate account of the dogged and, at times superhuman efforts, of Russian scientists, engineers and cosmonauts, to probe the unknown regions of [space], and to push the frontiers of science beyond the limits of human phantasy'. Concentrates on technical details, largely ignoring politics. Includes 30 pages of tables of rocket launches and orbital parameters of satellites and other spacecraft.

840 Riabchikov, Evgeny, (Nikolai P. Kamanin, ed., Guy Daniels, trans.)
RUSSIANS IN SPACE
New York: Doubleday, 1971; London: Weidenfeld & Nicolson, 1972, ISBN 0-297-00355-0. 300 + 16p; illus.

An official account of the Soviet space programme, for the general reader, prepared by the Novosti Press Agency Publishing House in Moscow.

841 Smolders, Peter, (Marian Powell, trans.)
SOVIETS IN SPACE (rev. ed.)
Guildford: Lutterworth Press, 1973, ISBN 0-7188-1990-X; New York: Taplinger, 1974, ISBN 0-8008-7340-8 (trans. of Dutch ed. 1971). 285 + 40p; illus (some col.), diags. Index (3).

The story of Soviet involvement with space, which concentrates more on technical than political matters, covering details such as diet and training of cosmonauts.

842 Stoiko, Michael
SOVIET ROCKETRY: THE FIRST DECADE OF ACHIEVEMENT (UK)/SOVIET ROCKETRY: PAST PRESENT AND FUTURE (US)
Newton Abbot: David & Charles, 1971, ISBN 0-7153-5308-X; New York: Holt, 1970, ISBN 0-03-081865-6. 272p; illus, diags. Bibl (1), index (2).

Beginning with a chapter on Konstantin Tsiolkovsky, the early rocket pioneer, this book describes the progress in Soviet rocketry which led to the launch of Sputnik 1 in October 1957. The main part of the book, however, concentrates on the following decade, dealing with launch vehicles, Earth satellites, the cosmonauts, and manned space missions. The author, an American, states that he 'minutely researched . . . official and authoritative sources'.

843 Vladimirov, Leonid, (David Floyd, trans.)
THE RUSSIAN SPACE BLUFF
London: Tom Stacey, 1971, ISBN 0-85468-023-3; New York: Dial, 1973, ISBN 0-8037-7539-3. 192 + 8p; illus. Index (2).

A Russian engineer who defected to the West disputes the idea that there was a space race, arguing that the US was always much further ahead technologically than the Soviet Union. The myth, as he describes it, continued after the Apollo 11 Moon landing, when Western journalists blindly accepted Soviet propaganda and published alarmist reports of their achievements. Not an anti-Russian book; the author expresses great admiration for the engineers who struggled to overcome the barriers and difficulties posed by the bureaucratic, secretive Soviet methods of running their space installations and the inefficiency of Soviet industry.

Other topics

844 Clark, John D.
IGNITION! AN INFORMAL HISTORY OF LIQUID ROCKET PROPELLANTS
New Brunswick, NJ: Rutgers, 1972, ISBN 0-8135-0725-1. 214p. Glos (2), index (3).

This book covers an unusual topic, describing the development of liquid propellants, which ran parallel to that of the rockets for which they were designed. Not over-technical, the book is intended for the interested lay reader, as well as the professional engineer in the rocket business.

845 Ezell, Edward Clinton and Linda Neuman Ezell
THE PARTNERSHIP: A HISTORY OF THE APOLLO-SOYUZ TEST PROJECT
Washington, DC: NASA, 1978, NASA SP-4209 pbk. 560p; illus (some col.), diags. Bibl (3), index (3).

A detailed, largely non-technical account, though not intended for casual reading, which represents the official US version of the historic link-up in space in 1975. Describes the backdrop of US-Soviet rivalry, and the idea, development and eventual success of the project.

846 Froehlich, Walter
APOLLO SOYUZ
Washington, DC: NASA, 1976, NASA EP-109. 131p; illus & diags (mostly col.).

A brief colourful account of the joint US-Soviet space mission which took place in 1975. Chapter headings are in Russian as well as English. The appendices provide information on the scientific experiments carried out.

847 Huzel, Dieter K.
PEENEMÜNDE TO CANAVERAL
Westport, CT: Greenwood Press, 1981, ISBN 0-313-22928-7 (reprint of 1st ed. 1962). 247 + 4p; illus. Index (2).

A popular account of the development of German rocketry up to the migration of Wernher von Braun and other rocket scientists to America and the start of their work at Cape Canaveral. Includes a chronology of German rocketry.

848 Lebedev, Lev and Alexander Romanov, (Leonard Stoklitsky, trans.)
RENDEZVOUS IN SPACE: SOYUZ-APOLLO
Moscow, Progress Publishers, 1979. 209p; illus (some col.), diags.

The Soviet version of the historic space mission, presented in a journalistic day-by-day account in which the authors take turns to describe the events and report the comments of the people involved.

849 Ordway, Frederick I., III and Mitchell R. Sharpe
THE ROCKET TEAM
New York: Crowell, 1979, ISBN 0-690-01656-5; London: Heinemann, 1979, ISBN 0-434-55300-X; Cambridge, MA: MIT Press, 1982; ISBN 0-262-65013-4 pbk. 462 + 32p; illus. Bibl (3), index (3).

The story of the German rocket pioneers, among whom Wernher von Braun is the best known, who began as enthusiastic amateurs in the late 1920s, but were turned through the German war effort 'into one of history's most sophisticated and intensely supported research and development teams'. Transferred to the US, 'the nucleus of the team helped develop and guide the technology that carried man to the Moon'. Incorporates the results of much original research, including about 200 direct interviews, and, through lengthy persuasion, much previously classified material. Includes a 34-page bibliography.

850 Powers, Robert M.
PLANETARY ENCOUNTERS
Harrisburg, PA: Stackpole, 1978, ISBN 0-8117-1270-2; London: Sidgwick & Jackson, 1982, ISBN 0-283-98764-2. 368p; illus, diags. Bibl (1), glos (1), index (2).

The history and possible future of unmanned space flight, told in a readable style for the general reader, in this book which won the Space Writers' award. Begins with the development of the idea of space flight, then goes on to discuss the various spacecraft, up to the beginnings of the Voyager missions, and summarize their findings. Also considers the possibility of interstellar probes.

851 Winter, Frank H.
PRELUDE TO THE SPACE AGE: THE ROCKET SOCIETIES, 1924–1940
Washington, DC: Smithsonian Institution Press (National Air & Space Museum), 1983, pbk. 222p; illus. Q

This book 'provides, for the first time, a prime reference document embracing the activities of all known rocket societies in the period 1924–1940'. It is also 'a tribute to the dedicated enthusiasts who struggled against the odds to bring their dreams to reality'. Following an account of the societies, by country, there is a 51-page photographic section.

Biography

852 Aldrin, Edwin E. and Wayne Warga
RETURN TO EARTH
New York: Random House, 1973, ISBN 0-394-48832-6; Toronto: Bantam, 1974, ISBN 0-552-68486-4 pbk. 342 + 16p; illus.

The autobiography of the second man to set foot on the Moon.

853 Bergaust, Erik
WERNHER VON BRAUN
Washington, DC: National Space Institute, 1976, ISBN 0-917680-01-4; and Harrisburg, PA: Stackpole, 1976, ISBN 0-8117-2098-5. 589p; illus. Bibl (2), index (3).

An authorized biography, which includes much reported conversation, of the man who contributed greatly to the success of the US space programme, following his wartime work on the German V weapons.

*854 Cipriano, Anthony J.
AMERICA'S JOURNEYS INTO SPACE: THE ASTRONAUTS OF THE UNITED STATES
New York: Wanderer, 1979, ISBN 0-671-33033-0 pbk. 212p.

Brief biographies of the astronauts involved in the Mercury, Gemini, Apollo, Skylab, and Apollo-Soyuz projects, with short mission summaries. (From *Spaceflight* May 1980, p 228).

855 Collins, Michael
CARRYING THE FIRE: AN ASTRONAUT'S JOURNEYS
New York: Farrar, 1974, ISBN 0-374-11917-1; London: W.H. Allen, 1975, ISBN 0-491-01603-4. 483 + 32p; illus, diags.

The author's own account of his career (not written with the assistance of a professional writer), from his selection for astronaut training to his two space missions, Gemini 10 and the historic Apollo 11. Collins was the man who remained in orbit while Armstrong and Aldrin landed on the Moon.

856 Cunningham, Walter, (assisted by Mickey Herskowitz)
THE ALL-AMERICAN BOYS
New York: Macmillan, 1977, ISBN 0-02-529240-4. 321 + 8p; illus. Index (2).

An account of Cunningham's career as an astronaut, which also covers the wider aspects of the US space programme, focusing on the other astronauts involved.

857 Grissom, Betty and Henry Still
STARFALL
New York: Crowell, 1974, ISBN 0-690-00473-7. 276 + 16p; illus. Bibl (2), index (2).

A biography of astronaut Virgil ('Gus') Grissom, whose career, along with those of Roger Chaffee and Ed White, ended tragically in January 1967 when a fire broke out in their Apollo spacecraft at the Kennedy Space Center. (*See also* 822).

858 Hooper, Gordon R.
THE COSMONAUTS 1977
Frinton on Sea: G.R. Hooper, 1977, pbk. 75p.

This privately printed booklet presents a record of the lives and careers of about forty known Soviet cosmonauts, and a number of unconfirmed ones.

859 Irwin, James B., (with William A. Emerson)
TO RULE THE NIGHT: THE DISCOVERY VOYAGE OF ASTRONAUT JIM IRWIN
Nashville, TN: Holman, 1973, ISBN 0-87981-024-6; London: Hodder & Stoughton, 1974, ISBN 0-340-18755-7; Nashville, TN: Broadman, 1982, ISBN 0-8054-7227-4. 251 + 16p; illus.

A novelistic account of Irwin's career, including his voyage to the Moon as part of the Apollo 15 crew. Tells his life story, with his final rediscovery of the Christian faith, enhanced by his experiences as an astronaut.

860 Irwin, Mary and Madalene Harris
THE MOON IS NOT ENOUGH
Grand Rapids, MI: Zondervan, 1978, ISBN 0-310-37050-7; Basingstoke: Pickering & Inglis, 1979, ISBN 0-7208-0446-9. 176 + 16p; illus.

An account of Mary Irwin's life with her husband, astronaut James Irwin, describing the effects of the space missions on their lives, their troubles, and their eventual spiritual awakening. An often melodramatic blend of romance and religion.

861 Killian, James R., jr.
SPUTNIK, SCIENTISTS AND EISENHOWER: A MEMOIR OF THE FIRST SPECIAL ASSISTANT TO THE PRESIDENT FOR SCIENCE AND TECHNOLOGY
Cambridge, MA: MIT Press, 1977, ISBN 0-262-11066-0. 315p; illus. Bibl (1), index (2).

A political autobiography, concentrating on the Sputnik affair and the reactions provoked in US government circles.

862 O'Leary, Brian
THE MAKING OF AN EX-ASTRONAUT
Boston: Houghton Mifflin, 1970; London: Michael Joseph, 1971, ISBN 0-7181-0849-3. 243 + 8p; illus, diags. Index (2).

The story of O'Leary's career as an astronaut, from his initial application to his ultimate disillusionment with the bureaucratic machinery which he found stifling to the quest for scientific knowledge. The test-pilot orientation of the space programme isolated him from the mainsteam of scientific research.

863 Scott, John M.
COUNTDOWN TO ENCOUNTER: VON BRAUN AND THE ASTRONAUTS
Huntington, IN: Our Sunday Visitor, 1979, ISBN 0-87973-630-5 pbk. 144p; illus.

The story of von Braun and the US space programme up to the Apollo 8 unmanned Moon landing, interlaced with religious moralizing and scriptural quotations.

*864 Sharpless, Jack
THE EARTHBOUND OBSERVER: A PERSONAL LOOK AT THE PEOPLE OF NASA AND THE SPACE SHUTTLE EFFORT
Michigan City, IN: Sharpless Corporation, 1983, ISBN 0-9610696-0-0. 89p; illus.

865 Wolfe, Tom
THE RIGHT STUFF
New York: Farrar, 1979, ISBN 0-374-25032-4; London: Jonathan Cape, 1979, ISBN 0-224-01443-9; New York and London: Bantam, 1981, ISBN 0-553-20700-8 pbk. 436p.

The story of the seven project Mercury astronauts, Scott Carpenter, Gordon Cooper, John Glenn, Virgil Grissom, Walter Schirra, Alan Shepard and Deke Slayton, following an account of the early rocket test flights, notably by Chuck Yeager. Written in an evocative, often incantatory style, the author captures the feelings of the astronauts and the mood of the time, as he describes the preparations, the first flights, the public response and the personal lives of the men and their wives. A film version of this book was released in 1983.

Space Science

Other works dealing with space science topics, particularly planetary and lunar science, may be found throughout Part 6. The Solar System. For books covering investigations in wavelengths outside the visible part of the electromagnetic spectrum, carried out beyond the Earth's atmosphere, *see* (520–522), (524) and (525).

General

866 Cornell, James and Paul Gorenstein (ed.)
ASTRONOMY FROM SPACE: SPUTNIK TO SPACE SHUTTLE
Cambridge, MA: MIT Press, 1983, ISBN 0-262-03097-7; 1985, ISBN 0-262-53061-9 pbk. 248p; illus, diags. Bibl (x), index (2).

A collection of ten articles for the scientist and interested lay reader, based on a series of lectures given to celebrate the 25th anniversary of the launch of Sputnik 1. Presents an up-to-date overview of the progress of space astronomy, with coverage of the geology of the inner planets, exploration of the Moon, the Voyager results, the Sun, the ultraviolet sky, x-ray astronomy, the Galaxy, beyond the Galaxy, the future of space astronomy, and the Earth from space.

867 French, Bevan, M. and Stephen P. Maran (ed.)
A MEETING WITH THE UNIVERSE: SCIENCE DISCOVERIES FROM THE SPACE PROGRAM
Washington, DC: NASA, 1981, NASA EP-177 pbk. 221p; illus

(mostly col.). Index (2). Q

A survey of the progress in space science for the general reader and younger space enthusiast. Covers planetary exploration, the Sun and its influence on the Earth, astronomy and astrophysics, space medicine, and exobiology, concluding with a rather over-optimistic chapter on future space programmes. Includes many excellent illustrations.

868 Grey, Jerry, Peter Downey and Bruce Davis (ed.)
SPACE: A RESOURCE FOR EARTH
New York: American Institute of Aeronautics and Astronautics, 1977, pbk. 67p; illus (some col.), diags. Q

The purpose of this book is 'to identify and document the many applications of space systems to improving the quality of life on Earth'. It is also a sourcebook of information on the history, uses and impact on civilization of space systems, as well as technical aspects. Discusses communications stellite systems, navigation satellite systems, land observations, sea and maritime observations, atmospheric observations, new potentials: space processing, life sciences and space-based solar power.

869 Hanle, Paul A. and Von Del Chamberlain (ed.)
SPACE SCIENCE COMES OF AGE: PERSPECTIVES IN THE HISTORY OF THE SPACE SCIENCES
Washington, DC: Smithsonian Institution Press (National Air & Space Museum), 1981, ISBN 0-87474-508-X; 0-87474-507-1 pbk. 194p; illus, diags. Bibl (x), index (1). Q

The proceedings of a symposium on the history of space science held in 1979, intended for the student and the reader with some scientific background. After 1957 scientists were freed from the limitations imposed by their earthbound location; it became possible to transport observatories, probes and other scientific hardware into space. This survey covers the sciences involved, and notable achievements, as well as considering the technological developments that made them possible.

870 Lawton, A.T.
A WINDOW IN THE SKY: ASTRONOMY FROM BEYOND THE EARTH'S ATMOSPHERE
Newton Abbot: David & Charles, 1979, ISBN 0-7153-7757-4; New York: Pergamon, 1979, ISBN 0-08-024663-X. 223p; illus. Index (2).

An account of the techniques of, and justifications for placing astronomical equipment in Earth orbit, for the interested lay reader. Considers the merits of possible observatory sites, discusses the progress in the newer branches of astronomy, and technological developments in observing instrumentation. Also looks at cosmology, and discusses possible future developments.

871 Lunan, Duncan
NEW WORLDS FOR OLD
Newton Abbot: Westbridge Books, 1979, ISBN 0-7153-7772-8; New York: Morrow, 1979, ISBN 0-688-03486-1. 268 + 8p; col. illus, diags. Bibl (3), index (2).

A survey of manned and unmanned exploration of the Solar System, covering methods and findings. Aims to emphasize the importance of exploration and the need for it to continue. Includes some chapters contributed by other writers. (*See also* 918).

Specific topics

872 Fishlock, David (ed.)
A GUIDE TO EARTH SATELLITES
London: Macdonald, 1971, ISBN 0-356-03867-X; New York: Elsevier, 1971, ISBN 0-444-19581-5. 159p; illus, diags. Index (1).

An introduction for the lay reader, in which eight articles discuss the technology of Earth satellites; the different types of satellite: communications, meteorological, Earth resources, navigational, and scientific; the military uses of satellites; concluding with 'advanced satellite concepts'.

873 Harper, Dorothy
EYE IN THE SKY
Montreal: Multiscience, 1977, ISBN 0-919868-01-0 pbk (reprint of 1st ed. 1976). 164p; illus (mostly col.), diags. Bibl (1), Index (1).

This introduction for the interested general reader describes the technology of remote sensing, and explains the importance of this use of satellites in fields such as agriculture and weather forecasting.

874 Lundquist, Charles A. (ed.)
SKYLAB'S ASTRONOMY AND SPACE SCIENCES
Washington, DC: NASA, 1979, NASA SP-404. 122p; illus & diags (mostly col.). Index (2).

A highly illustrated account of experiments carried out aboard the Skylab space station. Covers studies of stellar and galactic astronomy, interplanetary dust, Comet Kohoutek, energetic particles, Earth's atmosphere, and the orbital environment. An appendix summarizes the experiments in tabular form.

*875 McRoberts, Joseph J.
SPACE TELESCOPE
Washington, DC: NASA, 1982, NASA EP-166 pbk. 64p; illus.

Traces the historical and technological development of the space telescope to be launched with the Space Shuttle in [1986]. (From US GPO *Subject bibliography* SB-297, p 16).

*876 Moulton, Robert R.
FIRST TO FLY
Minneapolis, MN: Lerner, 1983, ISBN 0-8225-1576-8. 120p.

The behind-the-scenes story of Todd Nelson, the first participant in the Shuttle Student Involvement Project, whose 'Insect Flight Motion Study' was carried out aboard Columbia on mission STS-3. Aimed primarily at younger readers. (From *Space Age Times* May/June 1985, p 28).

877 NASA Lyndon B. Johnson Space Center
SKYLAB EXPLORES THE EARTH
Washington, DC: NASA, 1977, NASA SP-380. 517p; illus (mostly col.), diags. Bibl (x), glos (2). Q

A collection of 18 articles reporting the results of the three manned Skylab missions (1973–74), covering those studies which were devoted to the Earth. Technical in places, but heavily illustrated.

878 Page, Lou Williams and Thornton
APOLLO-SOYUZ PAMPHLETS 1–9
Washington, DC: NASA, 1977, NASA EP-133 to 144 pbk. Illus & diags (mostly col.). Glos (2), bibl (1) (most vols.).

Intended primarily for students and teachers of space science, these publications summarize the results of the joint US-Soviet space mission in 1975. Individual titles are: 1. *The flight* (57p), 2. *X-rays, gamma rays* (61p), 3. *Sun, stars, in between* (54p), 4. *Gravitational field* (30p), 5. *The Earth from orbit* (57p), 6. *Cosmic ray dosage* (36p), 7. *Biology in zero-G* (49p), 8. *Zero-G technology* (59p), 9. *General science* (76p).

879 Porter, Richard W.
THE VERSATILE SATELLITE
Oxford: Oxford University Press, 1977, ISBN 0-19-885104-9. 173p; illus (some col.), diags. Index (1).

A non-technical introduction to Earth satellites, which covers their history, and provides background information on communications, meteorology and navigation to show the importance of satellites in these areas. The author also deals with biological and agricultural monitoring, scientific experiments and astronomy in orbit, and manned satellites.

880 Richason, Benjamin F., jr. (ed.)
INTRODUCTION TO REMOTE SENSING OF THE ENVIRONMENT (2nd ed.)
Dubuque, IA: Kendall/Hunt, 1983, ISBN 0-8403-2834-6. 582p; illus (some col.), diags. Glos (2), bibl (x, 2), index (2).

A textbook for the non-science undergraduate which requires only a basic knowledge of physics. It is divided into two parts: 1. fundamentals, and 2. applications. There is a separate laboratory manual.

881 Summerlin, Lee B.
SKYLAB: CLASSROOM IN SPACE
Washington, DC: NASA, 1977, NASA SP-401. 182p; illus & diags (mostly col.). Index (2). Q

A highly illustrated, non-technical account of the Skylab Student Project, which allowed selected US high school students to submit proposals for experiments and design them. The book provides background information, describes the experiments and the results achieved. Covers also the science demonstrations carried out by astronauts to show the effects of weightlessness.

*882 van Nuss, Wayne D. and William W. Heusner
SPACE FLIGHT RESEARCH RELEVANT TO HEALTH, PHYSICAL EDUCATION AND RECREATION WITH PARTICULAR REFERENCE TO SKYLAB'S LIFE SCIENCE EXPERIMENTS
Washington, DC: NASA, 1979, NASA EP-148. 60p; illus, diags. Glos, index.

An introduction to the space programme for the professions named in the title, in particular to research relevant to the health-related fields of education, especially that deriving from the long-duration stays of astronauts aboard Skylab. (From Blashfield (15), p 118).

Spacecraft

General

883 Belew, Leland F. and Ernst Stuhlinger
SKYLAB: A GUIDEBOOK
Washington, DC: NASA, 1973, NASA EP-107 pbk. 245p; illus. Bibl (1), glos (2).

A brief overview of Skylab, its history, design, operation, and its programme of experiments and observations.

884 Belew, Leland F. (ed.)
SKYLAB: OUR FIRST SPACE STATION
Washington, DC: NASA, 1977, NASA SP-400. 164p; illus & diags (mostly col.). Index (2). Q

A highly illustrated account of the development, launch and operation of Skylab, intended for the general reader. Covers the technology and the people involved, with a brief summary of the astronomical results. This is the introductory volume to a series of popular reports on Skylab (*see also* 627, 874 and 881).

885 Cooper, Henry S.F.
A HOUSE IN SPACE
New York: Holt, 1976, ISBN 0-03-016686-1; London: Angus & Robertson, 1977, ISBN 0-207-95722-3; London: Granada (Panther), 1978, ISBN 0-586-04762-X pbk. 184 + 16p; illus, diags.

A journalistic account of the Skylab missions, told from the point of view of the astronauts who lived and worked on board. The text runs in a single sequence, with no division into chapters, and no contents page or index.

886 Froehlich, Walter
SPACELAB: AN INTERNATIONAL SHORT-STAY ORBITING LABORATORY
Washington, DC: NASA, 1983, NASA EP-165 pbk. 77p; illus & diags (mostly col.).

This brief, highly illustrated summary describes what Spacelab is (an Earth-orbiting laboratory carried by the Space Shuttle), the astronauts involved, the experiments to be carried out, a typical day aboard, ending with a chapter on its birth, impact and future.

887 Harrison, Harry and Malcolm Edwards
SPACECRAFT IN FACT AND FICTION
London: Orbis, 1979, ISBN 0-85613-192-X; New York: Exeter Books,

1979, ISBN 0-89673-019-0. 128p; illus (some col.). Bibl (1), index (1). Q

A pictorial history, illustrated with photographs and artwork. The brief, but interesting text covers fanciful early speculation, the development of rocketry, science fiction of the 1930s and 1940s, the space age, speculation on interstellar travel — fictional and serious, and recent cinema and TV.

888 Heaps, Leo
OPERATION MORNING LIGHT: TERROR IN OUR SKIES — THE TRUE STORY OF COSMOS 954
New York and London: Paddington Press, 1979, ISBN 0-448-22425-9 (US), 0-7092-0323-3 (UK); New York: Ballantine, 1979, ISBN 0-345-28101-2 pbk. 208p; illus. Glos (1), bibl (2), index (2).

A dramatized, journalistic account of the fall of the Soviet satellite, Cosmos 954, into Canada's North-West Territory, and the subsequent search for radioactive debris, leading to what the author claims was a government cover-up.

*889 Morrison, Colin A.
VOYAGE INTO THE UNKNOWN: THE SEARCH FOR AND RECOVERY OF COSMOS 954
Stittsville, Ontario: Canada's Wings, 1982, ISBN 0-920002-20-X. 154 + 32p; illus. Bibl (3), index.

*890 Kennedy, Gregory P. (compiler)
ROCKETS, MISSILES AND SPACECRAFT OF THE NATIONAL AIR AND SPACE MUSEUM (rev. ed.)
Washington, DC: Smithsonian Institution, 1983, pbk. 165p; illus.

Revised up to the summer of 1983, this catalogue illustrates the major space artefacts on display in the Museum. Exhibits range from 19th century Congreve rockets to detailed models of the Hubble Space Telescope. (From *S&T* June 1984, p 256).

*891 Malone, Robert and Jean Claude Suares
ROCKETSHIP
New York: Harper & Row, 1977, ISBN 0-06-012851-8 pbk. 127p; illus.

This picture book is a light-hearted blend of illustrations from science fiction (including movies and comic strips) and photographs of actual space exploration. (From *S&T* Dec. 1977, p 519).

892 REPORT ON THE SCIENTIFIC SATELLITES OF THE EUROPEAN SPACE AGENCY
Paris: European Space Agency, annual, pbk. About 110p; illus (some

col.), diags.

This survey for the professional or enthusiast provides brief, illustrated descriptions of satellites launched, including Spacelab payloads and projects under development.

ESA also produces a number of pamphlets for the general public giving a brief, colourful introduction to European space projects; examples include *Giotto* (Paris: ESA, 1981) and *Exosat* (Paris: ESA, 1982).

893 Shapland, David and Michael Rycroft
SPACELAB: RESEARCH IN EARTH ORBIT
Cambridge: Cambridge University Press, 1984, ISBN 0-521-26077-9. 192p; illus (some col.), diags. Index (1).

This profusely illustrated book describes briefly this European (with US co-operation) achievement: the development of a short-stay orbiting laboratory and its successful first launch on the Space Shuttle in November 1983. Covers also the experiments carried out, and some of the early results.

*894 Zimmerman, Howard
SPACESHIPS (2nd ed.)
New York: Starlog, 1980, ISBN 0-931064-23-6 pbk.

Space Shuttle

895 Allaway, Howard
THE SPACE SHUTTLE AT WORK
Washington, DC: NASA, 1979, NASA SP-432 & NASA EP-156 pbk. 76p; illus & diags (mostly col.). Index (1).

A brief, popular account of the Shuttle, which includes a fold-out cutaway illustration of the craft.

896 Allen, Joseph P. and Russell Martin
ENTERING SPACE: AN ASTRONAUT'S ODYSSEY
New York: Stewart, Tabori & Chang, 1984, ISBN 0-941434-53-2; London: Orbis, 1984, ISBN 0-85613-735-9. 223p; col. illus. Bibl (1), index (2).

A brief account of astronaut Allen's first Shuttle flight on Columbia in November 1982, enhanced with many excellent, well reproduced photographs. An extra chapter discusses the manned and unmanned exploration of the Solar System.

897 Baker, David
SPACE SHUTTLE
London: New Cavendish, 1979, ISBN 0-904568-23-7; 0-904568-22-9 pbk; New York: Crown, 1979, ISBN 0-517-53885-7; 0-517-53649-8 pbk. 72p; illus (mostly col.). Index (1). Q

This account, written before the Shuttle's first flight, is aimed primarily at younger readers, but among the mass of illustrations the densely packed text provides much information on the development and testing of the Shuttle. Includes a section, "For the modeller — adapting the basic Airfix Shuttle kit" by Matt Irvine.

898 Chant, Christopher
SPACE SHUTTLE
London: McLaren Publishing (Hong Kong) Ltd., 1984, ISBN 0-947889-00-0. 128p; illus (some col.), diags. Q

A brief, highly illustrated guide to the Shuttle and the 14 missions carried out up to the time of writing.

899 COUNTDOWN
(Main Stage Publications, PO Box 216, Athens, OH 45701, USA), 1983–, monthly. 24–28p; illus. Ads. Q

Began as 'the world's only monthly devoted exclusively to coverage of the Space Shuttle' and included in this section before the arrival of a letter from the publisher indicating that the magazine is growing (September 1985), and that while coverage is focused on the Space Shuttle (and the Space Station), it has broadened to include other areas of space exploration. (*See also* 787).

900 Dooling, Dave (ed.)
SHUTTLE TO THE NEXT SPACE AGE
New York: American Institute of Aeronautics and Astronautics, 1979, pbk. 131p; illus, diags. Q

A collection of papers from a two-day seminar held in Huntsville, Alabama in 1979, covering planned future space activities involving the Space Shuttle, with additional material about NASA, and European and Soviet programmes.

901 Gatland, Kenneth, Mark Hewish and Pearce Wright
THE SPACE SHUTTLE HANDBOOK
London: Hamlyn, 1979, ISBN 0-600-38418-7. 97p; illus & diags (mostly col.). Index (1). Q

A highly illustrated guide to the development of the Shuttle and its uses, including descriptions of missions to be carried out and the conditions in which the astronauts will work.

902 Joels, Kerry Mark, Gregory P. Kennedy and David Larkin
THE SPACE SHUTTLE OPERATOR'S MANUAL
New York: Ballantine, 1982, ISBN 0-345-30321-0 pbk; London: Macmillan (Papermac), 1983, ISBN 0-333-35109-9 pbk. 160 + 3 (fold-out)p; illus (some col.), diags. Glos (1). Q

The next best thing for those who would like to make the trip but probably never will, a compendium of all the information the would-be pilot or crew member is likely to need. Includes detailed plans of instrument consoles and other structural details. Written in the second person, the book covers everything from launch, through hygiene, working in space, emergency procedures, to re-entry and landing, with details of five Shuttle missions. The appendices provide technical details in lay person's terms, with many explanatory diagrams.

903 Kaplan, Marshall H.
SPACE SHUTTLE: AMERICA'S WINGS TO THE FUTURE
Fallbrook, CA: Aero Publishers, 1978, ISBN 0-8168-8450-1. 215p; illus (some col.), diags. Glos (1), index (1). Q

Written before the problems which delayed the first flight of the Shuttle occurred, this highly optimistic account traces the development of the Shuttle, and considers the benefits likely to be gained from it.

904 Kerrod, Robin
SPACE SHUTTLE
London: Multimedia Publications (Kola Books), 1984, ISBN 1-85106-009-X. 71p; col. illus. Q

A picture book which presents a brief account of the Shuttle.

905 Lewis, Richard S.
THE VOYAGES OF *COLUMBIA*: THE FIRST TRUE SPACESHIP
New York: Columbia University Press, 1984, ISBN 0-231-05924-8. 223p; illus, diags. Bibl (2), index (2). Q

Despite the title this is a general account of the Shuttle project, non-technical, with many interesting photographs and excerpts from ground to space conversations. The author, an experienced space writer, deals with the conception, development, testing and successful missions of the Shuttle, without omitting the problems overcome along the way. Includes a table summarizing the first nine missions.

906 Macknight, Nigel
SHUTTLE
Nottingham: Macknight International, 1984, ISBN 0-9509357-0-0 pbk. 98p; illus (mostly col.). Q

An attractively illustrated account written at Cape Canaveral by a young British author. Includes interviews with astronauts Robert Crippen and Mary Cleave, and a double-sided pull-out poster inside the back cover. A six (and one extra) -issue partwork derived from this book, titled *Shuttle Story*, was published in 1984–85. (*See also* 783).

907 NASA Lyndon B. Johnson Space Center
SPACE SHUTTLE
Washington, DC: NASA, 1976, NASA SP-407 pbk. 89p; illus & diags (mostly col.).

An early, highly illustrated, popular account of the Shuttle project.

*908 Powers, Robert M.
THE WORLD'S FIRST SPACESHIP: SHUTTLE
Harrisburg, PA: Stackpole, 1979, ISBN 0-8117-2112-4; New York: Warner, 1980, ISBN 0-446-95331-8 pbk. 255p; illus.

Describes Shuttle operations as they will take place in the near future (as of 1979), mixing in a little history and basic facts about Shuttle hardware. Covers such topics as launch and servicing facilities, Shuttle astronauts, orbital operations, including Spacelab, astronomy in space (mostly using the Space Telescope), and more distant future developments such as industries in space. (From *S&T* Feb. 1980, p 152).

909 Stockton, William and John Noble Wilford
SPACELINER: THE *NEW YORK TIMES* REPORT ON COLUMBIA'S VOYAGE INTO TOMORROW
New York: Times Books, 1981, ISBN 0-8129-0979-8. 183 + 23p; illus, diags. Index (3).

A readable, sometimes melodramatic account of the first Shuttle flight, including some background information on the development of the project, the uses of the Shuttle and its potential benefits to humanity.

The Future/Speculation

General

910 Adelman, Saul J. and Benjamin
BOUND FOR THE STARS
Englewood Cliffs, NJ: Prentice-Hall, 1981, ISBN 0-13-080390-1; 0-13-080382-0 pbk. 335p; illus. Index (3).

A survey of the problems of Solar System exploration and exploitation, and the possibility of interstellar travel. The authors suggest 'a practical course of action for the space enterprise. Our approach differs from the rosy unrealistic portrayals of the coming marvels of the Space Age painted in most popular books on space'.

911 Berry, Adrian
THE NEXT TEN THOUSAND YEARS: A VISION OF MAN'S FUTURE IN THE UNIVERSE
London: Jonathan Cape, 1974, ISBN 0-224-00966-4; New York: New American Library (Mentor), 1975, ISBN 0-451-61601-4 pbk. 224p. Glos (1), bibl (3), index (2).

A speculative look at what might be achieved in the future, assuming that economic and technological growth continue. Considers concepts such as lunar bases, terraforming Venus (adapting the planet to support human life), the construction of Dyson spheres (where the material of a star system is used to enclose the star), and other imaginative ideas.

912 Bova, Ben
THE HIGH ROAD
Boston: Houghton Mifflin, 1981, ISBN 0-395-31288-4; New York: Pocket Books, 1983, ISBN 0-671-45805-1 pbk. 289p. Index (2).

A science fiction writer and editor, who is a proponent of high technology discusses the future of human activity in space. Covers energy, industrialization, space settlements, military uses of space, with strong treatment of the political issues.

913 Calder, Nigel
SPACESHIPS OF THE MIND
London: BBC, 1978, ISBN 0-563-17477-3; New York: Viking, 1978, ISBN 0-670-66021-3; New York: Penguin, 1979, ISBN 0-14-005231-3 pbk. 144p; illus & diags (mostly col.). Index (1). Q

Produced to accompany a three-part series of programmes first broadcast by BBC TV in 1978, this book presents an optimistic account of our future in space, based on meetings with thinkers such as Freeman Dyson and Gerard O'Neill. Deals with a wide range of topics, from the problems of Earth, such as energy and population, to concepts such as space settlements, industry and transportation, as well as the possibility of interstellar travel.

914 Davies, Owen (ed.)
THE *OMNI* BOOK OF SPACE
New York: Kensington (Zebra), [1983], ISBN 0-8217-1295-6 pbk. 398p.

A collection of 33 articles reprinted from *Omni* magazine, 1979–82, mostly of a speculative nature.

*915 DeNevi, Don
TO THE EDGES OF THE UNIVERSE: SPACE EXPLORATION IN THE TWENTIETH CENTURY
Millbrae, CA: Celestial Arts, 1978, ISBN 0-89087-184-1; 0-89087-212-0 pbk. 214p; illus. Glos, bibl, index.

A non-technical exploration of the ideas developed both inside and outside the space programme for human travel outward into the universe. After a brief background survey, the author investigates methods of occupying space, colonizing nearby planets and exploring the distant reaches of the Solar System. (From Blashfield (15), p 56).

916 Glenn, Jerome Clayton and George S. Robinson
SPACE TREK: THE ENDLESS MIGRATION
Harrisburg, PA: Stackpole, 1978, ISBN 0-8117-1581-7; New York: Warner, 1980, ISBN 0-446-91122-4 pbk. 223p; illus, diags. Bibl (2), index (2).

This popular account of the possibility of human migration into space discusses the Space Shuttle, colonies in space, war in space, industrialization, and looks at the political factors involved, including a discussion of the organizations and groups pressing for action.

917 Grey, Jerry
BEACHHEADS IN SPACE: A BLUEPRINT FOR THE FUTURE
New York: Macmillan, 1983, ISBN 0-02-545590-7. 274p.

An optimistic survey of the current state of progress in space, and prospects for the future. The concept of beachheads (the coming development of permanent space stations) suggests a quasi-military attack on a hostile environment. An index might have been useful.

918 Lunan, Duncan
MAN AND THE PLANETS: THE RESOURCES OF THE SOLAR SYSTEM
Bath: Ashgrove Press, 1983, ISBN 0-906798-17-5. 306 + 12p; illus, diags. Bibl (3), index (2).

A follow-up to *New worlds for old* (871), which includes updating material to the end of 1981, but is largely about 'the prospects for future human life beyond the Earth, throughout our Solar System'. Covers transportation, space settlements, the exploitation of planetary resources, and other projects, from those currently proposed to those approaching the wildest realms of speculation.

919 Moore, Patrick and David Hardy
THE NEW CHALLENGE OF THE STARS (2nd ed.)
New York: Rand McNally, 1977, ISBN 0-528-83087-2; London: Mitchell Beazley, 1978, ISBN 0-85533-133-X. 63p; illus, diags (mostly col.). Index (1). Q

Speculation on the exploration and settlement of the Solar System and beyond, with paintings by Hardy and commentary by Moore. The text incorporates a basic introduction to astronomy, particularly that of the Solar System. Ends with several pages of more fanciful concepts, such as a photon starship, an asteroid ark and interstellar warfare.

920 Moore, Patrick
THE NEXT FIFTY YEARS IN SPACE
London: William Luscombe, 1976, ISBN 0-86002-033-9; New York: Taplinger, 1976, ISBN 0-8008-5528-0; 1978, 0-8008-5529-9 pbk. 144p; illus (mostly col.). Index (2). Q

Looks at the possible future of Solar System exploration, taking various aspects, e.g. satellites, probes to the inner planets, a lunar base, man beyond Mars, and considers their likely development. Includes colour artwork by Andrew Farmer.

921 Oberg, Alcestis R.
SPACEFARERS OF THE '80s AND '90s: THE NEXT THOUSAND PEOPLE IN SPACE
New York: Columbia University Press, 1985, ISBN 0-231-05906-X. 238p; illus. Bibl (2), index (2).

A book about humans in space: how they get there, why they go and the effect it has on them. Based on interviews with astronomers, scientists, engineers and technical experts, inside and outside

NASA. Topics covered include NASA selection procedures and training programmes, civilians and businesses in space, space stations and interplanetary travel.

*922 Oberg, James E.
THE NEW RACE FOR SPACE
Harrisburg, PA: Stackpole, 1984, ISBN 0-8117-2177-9 pbk. 210p; illus, diags.

An authoritative and comprehensive look at the future goals, programmes and hardware of American and Soviet space exploration. (From *S&T* April 1985, p 321).

923 O'Leary, Brian
THE FERTILE STARS
New York: Everest House, 1981, ISBN 0-89696-079-X. 132p; illus (some col.). Index (2).

This optimistic account of mankind's possible future in space, considers how earthly problems of energy and food shortage, for example, may be eased or solved by the exploitation of space. The author argues that the space solution is inevitable, if only because some crisis may render the Earth unfit for human habitation. Considers such topics as transportation, power from space, settlements, asteroid mining, and the political issues involved.

924 O'Leary, Brian
PROJECT SPACE STATION: PLANS FOR A PERMANENT MANNED SPACE CENTER
Harrisburg, PA: Stackpole, 1983, ISBN 0-8117-1701-1. 159p; illus, diags. Bibl (2), Index (2).

Intended for the general reader, this book presents the case for a permanent manned space station, covering the historical background, possible uses of the station, and likely future developments.

925 Pardoe, Geoffrey K.C.
THE FUTURE FOR SPACE TECHNOLOGY
London and Dover, NH: Frances Pinter, 1984, ISBN 0-86187-462-5. 206 + 8p; illus (some col.), diags. Index (2).

This book, aimed at a wide readership of those who have or should have an interest in space, puts the case for the continued exploration and use of space. Chapters cover: achievements to date, the philosophy for the future use of space, unmanned scientific satellites and probes, communication navigation and

Earth observation, manned space stations, large structures in Earth orbit, colonization, space transportation systems, space technology and the developing countries, military space systems, the organization of space programmes, practical implications for management, and the far future.

*926 Simpson, Theodore R. (ed.)
THE SPACE STATION: AN IDEA WHOSE TIME HAS COME
New York: Institute of Electrical and Electronics Engineers, 1985, ISBN 0-87942-182-7. 295p; illus. Index.

Sixteen chapters by such authors as Ronald Reagan, James Beggs (NASA administrator) and Harrison Schmidt (ex-astronaut and senator), discuss the role of NASA's projected space station. The book is divided into four parts, covering historical background, the decision to build the station, station concepts and uses, and the long-term potential. Seven appendices reprint important policy statements, memos, and speeches, including Kennedy's Apollo speech, Nixon's Space Shuttle statement, and Reagan's State of the Union address on the Space Station. (From *Astronomy* Aug. 1985, p 40).

927 Stine, G. Harry
SPACE POWER
New York: Ace, 1981, ISBN 0-441-77744-9 pbk. 234p; diags.

This book presents the case for the development of solar power satellites to help meet Earth's energy needs. Explores the technical and logistic problems, but in general extols the virtues of this space-based energy source, in an account for the general reader.

*928 Walker, Laurence C.
THE EXPLORATION AND COLONIZATION OF SPACE: LESSONS FROM HISTORY
College Station, TX: Center for Educational Research, 1980, ISBN 0-86599-005-0 pbk. 28p.

Commercial use of space

929 Kinsley, Michael E.
OUTER SPACE AND INNER SANCTUMS: GOVERNMENT, BUSINESS AND SATELLITE COMMUNICATIONS
New York: Wiley, 1976, ISBN 0-471-48060-6. 280p. Bibl (3), index (2).

The story of commercial satellite communications for all those with an interest in the subject.

930 Marsh, Peter
THE SPACE BUSINESS: A MANUAL ON THE COMMERCIAL USES OF SPACE
Harmondsworth: Penguin (Pelican), 1985, ISBN 0-14-022517-X pbk. 232p; diags. Bibl (3), index (3).

This survey of current and potential uses of space begins with a brief history of rocketry, then moves on to look at communication satellites, TV satellites, remote sensing, spy satellites, space war, industry in space, robots in space, then back to rockets to consider their uses and likely development. The book ends with a look at future possibilities.

931 SPACE BUSINESS NEWS
(Pasha Publications, Inc., 1401 Wilson Blvd., Suite 910, Arlington, VA 22209-2306, USA), 1983–, bi-weekly, ISSN 0738-9884. 8p (+ supplements); diags. Q

This expensive newsletter is aimed primarily at company executives with an interest in the commercial use of space. Coverage includes news of NASA, companies, products, contracts, and also non-US space involvement. At the time of writing (September 1985) the publisher is offering a 278-page report, *Space processing, products and profits 1983–1990*, free with subscriptions.

*932 Stine, G. Harry
THE SPACE ENTERPRISE
New York: Ace, 1980, ISBN 0-441-77756-2 pbk. 226p; diags.

933 Stine, G. Harry
THE THIRD INDUSTRIAL REVOLUTION
New York: Putnam, 1975, ISBN 0-399-11552-8. 192p; diags. Bibl (1), index (2).

An advocate of the commercial use of space discusses the political background, the technology and the potential benefits, in this account for the general reader.

Military use of space

934 Baker, David
THE SHAPE OF WARS TO COME
Cambridge: Patrick Stephens, 1981, ISBN 0-85059-483-9; London: Hamlyn, 1982, ISBN 0-600-20651-3 pbk; Briarcliff Manor, NY: Stein & Day, 1984, ISBN 0-8128-6221-X. 177p; illus (some col.), diags. Index (1).

This book describes the use of rockets in warfare, and the development of the technology which would allow the conduct of a war in space, particularly highlighting the development of surveillance, meteorological, and so-called killersats, with applications in terrestrial warfare. Considers also the possible military applications of the Space Shuttle, and the political factors which could lead to a war in space.

935 Barnaby, Frank
SPACE WEAPONS
London: Multimedia Publications (Kola Books), 1984, ISBN 1-85106-021-9. 72p; col. illus. Q

Basically a picture book which presents a brief survey of recent and possible future developments.

936 BEAM DEFENSE: AN ALTERNATIVE TO NUCLEAR DESTRUCTION
Fallbrook, CA: Aero Publishers, 1983, ISBN 0-8168-4138-1 pbk. 154 + 16p; illus (some col.), diags.

Produced by the 'scientific staff of the Fusion Energy Foundation', this is a popular account of proposed anti-missile satellite weapon systems using directed laser, particle, plasma or microwave beams.

A more detailed assessment of beam weapons for use in combat, in space and elsewhere, is given by Jeff Hecht in *Beam weapons: the next arms race* (New York: Plenum, 1984, ISBN 0-306-41546-1).

937 Bova, Ben
ASSURED SURVIVAL: PUTTING THE STAR WARS DEFENSE IN PERSPECTIVE
Boston: Houghton Mifflin, 1984, ISBN 0-395-36405-1. 343p. Bibl (2).

The author, better known as a science fiction writer and editor, who is an advocate of the proposed anti-missile satellite weapon systems, writes that the message of this book is: 'we have the means almost within our grasp to put an end to war'. An appendix provides information on the different types of space weapon.

938 Canan, James
WAR IN SPACE
New York: Harper & Row, 1982, ISBN 0-06-038022-5; New York: Berkeley, 1984, ISBN 0-425-06848-X pbk. 192p. Index (2).

This popular account by a journalist who has worked as an intelligence officer, concentrates on the political factors relating to space warfare. Much of the book is devoted to the military and political influences on technological projects, showing how the present situation developed.

*939 Graham, Daniel O.
HIGH FRONTIER
New York: Pinnacle (Tor Books), 1983, ISBN 0-523-48078-4 pbk. 314p.

*940 Graham, Daniel O.
THE NON-NUCLEAR DEFENSE OF CITIES: THE HIGH FRONTIER SPACE-BASED DEFENSE AGAINST ICBM ATTACK
Cambridge, MA: Abt Books, 1983, ISBN 0-89011-586-9; Lanham, MD: University Press of America, 1984, ISBN 0-8191-4101-1. 152p; illus.

941 Jasani, Bhupendra and Christopher Lee
COUNTDOWN TO SPACE WAR
London and Philadelphia: Taylor & Francis, 1984, ISBN 0-85066-261-2 pbk. 104 + 12p; illus (some col.), diags. Bibl (1), index (2).

This book, published under the auspices of SIPRI (the Stockholm International Peace Research Institute), attempts to bring to the general reader an understanding of the issues involved in the military use of space. The emphasis is on the technological side: how the present situation developed, and a survey of current systems and their potential uses.

A detailed exposition of the technological aspects of space militarization can be found in another SIPRI publication: Bhupendra Jasani (ed.), *Outer space: a new dimension of the arms race* (Taylor & Francis, 1982, ISBN 0-85066-231-1). *See also*, Bhupendra Jasani (ed.), *Space weapons: the arms control dilemma* (Taylor & Francis, 1984, ISBN 0-85066-262-1).

942 [Jasani, Bhupendra]
OUTER SPACE — BATTLEFIELD OF THE FUTURE?
London: Taylor & Francis, 1978, ISBN 0-85066-130-7. 202p; illus, diags. Bibl (2), glos (2), index (2).

Published under the auspices of SIPRI (*see* 941), this book is an attempt to spread information about the military use of space, particularly in connection with the development of satellite technology. Describes the various types of satellite, providing appendices of data on all launched satellites. Ends with a discussion of the use of satellites for the verification of arms control agreements. Not for the casual reader.

943 Karas, Thomas
THE NEW HIGH GROUND: SYSTEMS AND WEAPONS OF SPACE AGE WAR
New York: Simon & Schuster, 1983, ISBN 0-671-47025-6; 1984, 0-671-50591-2 pbk. 224 + 16p; illus. Index (3).

An account of US military involvement in space; politics, industry, technology, weapons and war, defence of Earth satellites, and the (mistaken) use of laser weapons. The author writes: 'although it is indeed time that we recognize the military usefulness of space technology, we need also to understand that space power is not going to provide us with a military superiority that will solve all our problems'.

944 MANNO, Jack
ARMING THE HEAVENS: THE HIDDEN MILITARY AGENDA FOR SPACE, 1945–1995
New York: Dodd, Mead, 1984, ISBN 0-396-08211-4; 0-396-08212-2 pbk. 245p. Bibl (3), index (3).

This book traces the development of the US and Soviet space programmes, beginning with 'the Nazi legacy', showing how military plans formed as the programmes proceeded. The author is an opponent of space militarization.

945 MILITARY SPACE
(Pasha Publications, Inc., 1401 Wilson Blvd., Suite 910, Arlington, VA 22209-2306, USA), bi-weekly, ISSN 0743-7897. 8p (+ supplements). Q

This expensive newsletter is aimed primarily at company executives with an interest in space-related military developments. Provides coverage of the political and technical aspects of military space programmes. At the time of writing (September 1985) the publisher is offering a 150-page report, *Guide to military space programs*, free with subscriptions.

946 Peebles, Curtis, (Kenneth Gatland, consultant)
BATTLE FOR SPACE
Poole: Blandford, 1983, ISBN 0-7137-1378-X; New York: Beaufort Books, 1983, ISBN 0-8253-0160-2. 192p; illus & diags (some col.). Glos (1), bibl (1), index (1).

This highly illustrated account of the military uses of space for the general reader incorporates some information which was declassified at the author's request. Covers the space systems (such as launch and delivery), the 'nuclear factor', satellites and anti-satellites, laser and particle beam weapons, and the political factors. Soviet coverage is extremely limited because of the difficulty of obtaining information.

947 Ritchie, David
SPACEWAR
New York: Atheneum, 1982, ISBN 0-689-11264-5; New York: New American Library (Plume), 1983, ISBN 0-452-25461-2 pbk. 224p; illus. Bibl (3), index (2).

A sensationalist account, by a professional science writer, of the build-up of weapons with the potential for use in space, and the likelihood of this happening. Considers the US and Soviet space programmes and their military implications. Examples of the chapter headings used give a flavour of the book: "The kiss of death", "Dynosoarski" and "Dragon in the sky".

948 Stine, G. Harry
CONFRONTATION IN SPACE
Englewood Cliffs, NJ: Prentice-Hall, 1981, ISBN 0-13-167437-4; 0-13-167429-3 pbk. 209 + 8p; illus, diags. Bibl (2), index (2).

Intended for the general reader, this book avoids the sensationalism and superficiality of some of its kind to present a sober account of space transportation systems, weapons, and the principles of warfare in space. Not intended to promote the military use of space, but by accepting it as a reality aims to promote the study of war as a means toward eliminating it.

Human settlements

949 Allaby, Michael and James Lovelock
THE GREENING OF MARS
London: André Deutsch, 1984, ISBN 0-233-97655-8; New York: St. Martins, 1984, ISBN 0-312-35024-4. 166p.

A natural history writer and an atmospheric chemist consider the possibility of terraforming (*see* 958) the planet Mars and setting up human habitations, with technology now in existence.

950 Brand, Stewart (ed.)
SPACE COLONIES
New York: Penguin, 1977, ISBN 0-14-004805-7 pbk. 160p; illus, diags. Bibl (2), index (2).

A collection of material most of which originally appeared in *The CoEvolution Quarterly*, a magazine concerned with the details of life, rather than moods, feelings and generalizations. Intended to present a realistic picture of the possibilities and practicalities of space colonies, the book's three sections consider Gerard O'Neill's proposals (*see* 959), views for and against, and space (technology, astronomy, politics). Living up to its claims, the book includes an article on passing bodily wastes in weightless conditions. Packed with information, this collection includes reviews of information sources, addresses, interviews and personal accounts.

951 Gale, William A. (ed.)
LIFE IN THE UNIVERSE: THE ULTIMATE LIMITS TO GROWTH
Boulder, CO: Westview Press, 1979, ISBN 0-89158-378-5. 121p; diags. Bibl (x), index (1).

A collection of symposium papers, occasionally technical, which examine the prospects of human settlements beyond the Earth's atmosphere.

952 Golden, Frederic
COLONIES IN SPACE: THE NEXT GIANT STEP
New York: Harcourt Brace Jovanovich, 1977, ISBN 0-15-219400-2. 145p; illus, diags. Bibl (2), index (1).

Aimed primarily at younger readers, this is a straightforward non-technical account of the proposals made by Gerard K. O'Neill (959). Looks at the evolution of the idea of space colonies, and considers how they may be built, and what life in them may be like.

953 Heppenheimer, T.A.
COLONIES IN SPACE
Harrisburg, PA: Stackpole, 1977, ISBN 0-8117-0397-5; New York: Warner, 1978, ISBN 0-446-81581-0 pbk. 224 + 8p; illus (some col.), diags. Bibl (2), index (2).

An optimistic look at the possible establishment of human space habitats in the Earth-Moon system.

954 Heppenheimer, T.A.
TOWARD DISTANT SUNS
Harrisburg, PA: Stackpole, 1979, ISBN 0-8117-1578-7; New York: Fawcett (Columbine), 1980, ISBN 0-449-90035-5 pbk. 256 + 8p; illus (some col.), diags. Bibl (2), index (2).

This entertaining, optimistic account, by an aeronautical engineer, considers the possibility of near-space colonization carried out with a step by step approach, covering the technology, economics, politics and social factors.

955 Johnson, Richard D. and Charles Holbrow (ed.)
SPACE SETTLEMENTS: A DESIGN STUDY
Washington, DC: NASA, 1977, NASA SP-413 pbk. 185p; col. illus, diags. Bibl (x).

This study of near-space habitats is technical in places, but is meant for a readership beyond that of the aerospace community. Includes considerable background material.

*956 Marotta, Michael E.
SPACE COLONIZATION: AN ANNOTATED BIBLIOGRAPHY
Mason, MI: Loompanics, 1979, pbk. 32p.

A brief bibliography based on the belief that space colonization is now 'inevitable and that those who prepare now will ride a wave of the future'. Much of the material is basic to any coverage of contemporary aerospace planning. (From Blashfield (15), p 127).

957 Oberg, James E.
MISSION TO MARS: PLANS AND CONCEPTS FOR THE FIRST MANNED LANDING
Harrisburg, PA: Stackpole, 1982, ISBN 0-8117-0432-7; New York: New American Library (Meridian), 1983, ISBN 0-452-00655-4 pbk. 221p; illus, diags. Bibl (2), index (2).

An optimistic account of the possible manned exploration and settlement of Mars, which the author argues 'is technologically feasible as soon as we decide to do it, and can be justified on scientific, economic, social and political grounds'. Oberg expounds this thesis with infectious enthusiasm for the general reader.

958 Oberg, James E.
NEW EARTHS: TRANSFORMING OTHER PLANETS FOR HUMANITY
Harrisburg, PA: Stackpole, 1981, ISBN 0-8117-1007-6; New York: New American Library (Meridian), 1983, ISBN 0-452-00623-6 pbk. 283 + 4p; illus (some col.), diags. Glos (2), bibl (2), index (2).

Serious speculation on the concept of terraforming (adapting planets to make them capable of supporting Earth-based life), an idea once confined to the realms of science fiction. Provides a wealth of background detail on the formation of the Earth and the development of life, with further information on Mars and Venus, the prime candidates for modification. Considers climatic and biological factors, the technology of terraforming, and the practical and ethical questions.

959 O'Neill, Gerard K.
THE HIGH FRONTIER: HUMAN COLONIES IN SPACE
New York: Morrow, 1977, ISBN 0-688-03133-1; London: Jonathan Cape, 1977, ISBN 0-224-01406-4; London: Transworld (Corgi), 1978, ISBN 0-552-10827-7 pbk; Garden City, NY: Doubleday (Anchor Press), 1982, ISBN 0-385-18232-5 pbk. 288 + 24p; illus, diags. Bibl (2), index (2).

A landmark volume in the promotion of the concept of space settlements, taking the idea out of the realms of science fiction, and advocating action now that the necessary technology is available. Describes what living and working would be like in the huge 'islands', space communities complete with agriculture and industry, located at the stable L5 point of the Earth-Moon system. The book also deals with the author's public advocacy of this concept.

960 Robinson, George S.
LIVING IN OUTER SPACE
Washington, DC: Public Affairs Press, 1976, ISBN 0-8183-0243-7 pbk. 119p. Bibl (2), index (1).

Aimed at the general reader, this book is concerned more with the biological, social and psychological aspects of human activity in space than with the setting up of long-term space habitats.

*961 Tanner, Don and George Johnson
CITIES IN SPACE: OUR INCREDIBLE FUTURE . . . SCIENCE FICTION OR SCIENTIFIC FACT?
Eugene, OR: Harvest House, 1979, ISBN 0-89081-186-5 pbk. 157 + 4p; illus. Bibl.

Interstellar travel

962 Berry, Adrian
THE IRON SUN: CROSSING THE UNIVERSE THROUGH BLACK HOLES
London: Jonathan Cape, 1977, ISBN 0-224-01357-2; New York: Dutton, 1977, ISBN 0-525-13490-5; New York: Warner, 1978, ISBN 0-446-89796-5 pbk; London: Hodder & Stoughton (Coronet), 1979, ISBN 0-340-23231-5 pbk. 176p; diags. Glos (2), bibl (3), index (2).

A highly speculative account, convincingly presented for the general reader. Provides background information on physics, before considering the politics, economics and, above all, the technology of such a venture.

963 Gilfillan, Edward S., jr.
MIGRATION TO THE STARS: NEVER AGAIN ENOUGH PEOPLE
Washington, DC: Robert B. Luce, 1975, ISBN 0-88331-072-4. 226p.

This account, aimed at the general reader, is concerned mainly with the development of spacecraft and propulsion systems, but also covers the human aspects of living in space, and the problems likely to be encountered in interstellar travel.

964 Macvey, John W.
COLONIZING OTHER WORLDS: A FIELD MANUAL
Briarcliff Manor, NY: Stein & Day, 1984, ISBN 0-8128-2943-3. 216p; diags. Index (2).

Beginning with the premise that a fleet of ISTs (interstellar transports) will one day leave the Earth on a one-way colonizing expedition to the stars, the author discusses the problems likely to be faced on an earthlike planet; for example, indigenous life, communications, the search for minerals, agriculture, the atmosphere, law and government. He also considers other bodies in our Solar System as potential colony sites. Appendices cover cryogenic hibernation, finding earthlike planets and interstellar travel times.

965 Macvey, John W.
INTERSTELLAR TRAVEL: PAST PRESENT AND FUTURE
Briarcliff Manor, NY: Stein & Day, 1977, ISBN 0-8128-2278-1; New York: Avon, 1978, ISBN 0-380-41368-X pbk. 253p; diags. Bibl (2), index (2).

In the first part of this book the author, working on the assumption that the galaxy contains numerous intelligent inhabitants, considers possible means of propulsion across interstellar distances. This leads on to the second part in which, after briefly considering alien biochemistry, the author examines the evidence for extraterrestrial visits to Earth, past and present.

966 Macvey, John W.
SPACE WEAPONS/SPACE WAR
New York: Stein & Day, 1979, ISBN 0-8128-2579-9; London: New English Library, 1982, ISBN 0-450-05115-3 pbk. 245p; diags. Bibl (2), index (2).

Serious, though possibly a little alarmist speculation on the possibility of interstellar warfare involving the Earth. Considers the possibility of life in the universe, interstellar travel, defence against invasion, and the strategy, tactics and weapons of interstellar warfare. Includes a discussion of UFOs.

967 Nicolson, Iain
THE ROAD TO THE STARS
Newton Abbot: Westbridge Books, 1978, ISBN 0-7153-7618-7; New York: Morrow, 1978, ISBN 0-688-03336-9. 224p; illus (mostly col.), diags. Index (2).

The author begins with the assumption that interstellar travel is feasible, with journey times measured in decades rather than millenia. The book's three sections cover 1. the reasons for star flight, 2. basic principles, progress to date and possible technologies (in the longest section by far), and 3. sociological and psychological effects of interstellar travel.

968 Powers, Robert M.
THE COATTAILS OF GOD: THE ULTIMATE SPACEFLIGHT — THE TRIP TO THE STARS
New York: Warner, 1981, ISBN 0-446-51231-1; 1982, ISBN 0-446-90092-3 pbk. 288p; illus. Bibl (3), index (3).

A readable, wide-ranging survey of the factors involved in planning an interstellar space flight. The final chapters describe the flight of a hypothetical future starship. Includes an unusually long (27 pages) bibliography for a popular work.

Author Index

Numbers refer to entries. The letter 'n' following a number indicates an author mentioned only in the notes of the entry concerned; the letter 't' indicates a translator. Note that names beginning 'Mac' and 'Mc' are listed as spelt.

Title Index: Books

Numbers refer to entries. The letter 'n' following a number indicates a title mentioned only in the notes of the entry concerned, or the title of a previous edition. Some of the longer titles have been shortened.

Title Index: Periodicals

Numbers refer to entries. The letter 'n' following a number indicates a title mentioned only in the notes of the entry concerned, or a previous title. Annual publications are not listed here.

Subject Index

Numbers refer to entries. The letter 'n' following a number indicates an item mentioned only in the notes of the entry concerned.